THE SCIENCE OF REINCARNATION

BOB GOOD

Copyright © 2012 Bob Good

All rights reserved.

Cover art by Christopher Rosaferra

ISBN-10: 1478203854

EAN-13: 9781478203858

DEDICATION

This book is dedicated to Stephan Schwartz and Dean Radin, whose early support meant more than they knew and to all the scientists who do this work to expand the frontier of human awareness.

FOREWORD

This book is about the science of reincarnation. I write it not as a believer, but as someone with a scientific mind who thinks that the studies coming from laboratories around the world are so important and startling that they suggest a totally new understanding of ourselves and the world around us. These laboratories are discovering evidence that may provide the foundation for scientific proof that we reincarnate.

This book is not some spiritual, New Age treatise. It is not about psychology. I am not asking you to take anything on faith. This book is about probability and outcomes. I cite real scientific studies that are available for you to read. Based on the results of these studies, it seems probable to me that human beings reincarnate. After you read this book, I think you will agree with me.

The Science of Reincarnation will prove that reincarnation should not be accepted or rejected on faith, but examined scientifically, and it will show that the science of reincarnation meets all the criteria to be accepted as a science by the scientific community.

This book will also lay out how the science of reincarnation should be done. There have been advances in a number of fields; this book will show how they are interconnected, and how they collectively support the science of reincarnation. Beyond that, this book will show both statistically and geometrically that the mathematics supports the science of reincarnation.

The Science of Reincarnation will convince you that you have reincarnated before and will reincarnate again. It will do that by taking you from study to study, clearly and concisely showing you the logical and factual reasons this is so. All the information in this book will be documented so that you can check the sources. You can then make your own determination based on the facts, studies, and theories that have emerged in the last twenty-five years. Collectively, the results point to a new perspective on how we should see ourselves and what we may really be.

Finally, I submit to you that the science of reincarnation is already being done. This book is, in effect, a proposed syllabus for the science of reincarnation: how it should be taught, how it is being done, and how it should proceed, both individually in each of its disciplines, and collectively, as a body of scientific knowledge.

As you read this book, you can go to www.thescienceofreincarnation.com to watch videos of and hear lectures from each of the categories or disciplines of the scientists I discuss. The introductory five-minute video is about a child who you see between the ages of two and five. The child claims to be a World War II fighter pilot who died in the Pacific. The child's information was so specific that veterans of the unit say yes, that child was their comrade.

The website makes two points: one, that reincarnation should be considered a science and fits neatly into the emerging metaparadigm, and two, that the scientists in the various disciplines touching on reincarnation should work cohesively to achieve a better result.

The reason this is so important has to do with funding dollars. To do basic research in government or academia funding is critical. If reincarnation becomes an accepted science, funding dollars will flow to projects that could not be funded before. By working together, scientists will by consensus identify needed research areas. I am not talking about frivolous research, but supporting and expanding work already being done at institutions like Stanford, Princeton University, and The University of Virginia, to name a few.

This information is not static. This argument will be refined by scientists and writers who will do a better job than I have. The science will evolve, but the structure of this argument will be unchanged. Simply put, we are proving scientifically a human capability that transcends death. This book defines both that science and its paradigm.

But in a larger sense, the emergence of this science is part and parcel of the new evolving metaparadigm. This will do more than redefine our reality, it will redefine ourselves.

The science in this book treats us all equally, and as the case is built, that fact becomes inescapable. Whatever our differences of creed, color, race, or religion, all our hearts beat, and blood flows in all our veins, and we are more the same than different and more equal in death than unequal in life. The science is both karmic and democratic.

The ramifications of this thought have very real political and ideological consequences. If you don't think so, I will point out that China has passed a law that makes it illegal for the Dali Lama to reincarnate anywhere except on Chinese soil.

The fact that governmental legislation like this would even be considered, let alone passed into law, should indicate just how radical, in a geopolitical sense, the science of reincarnation case can be.

The end of all of this should be the creation of a 501(c)3 charitable foundation to advance the study of reincarnation. Funding should be distributed to academic institutions to specifically advance research into the science of reincarnation, which could be any of the disciplines discussed in this book, and more I haven't mentioned. And while I have never asked, nor have I met the people I refer to in this book, any of them would make good board members for this new foundation.

CHAPTER 1

CHILDREN WHO CLAIM PRIOR LIVES

There have been over 2500 cases catalogued by the University of Virginia over the course of 60 years of children who claim to have lived before. Typically, these children are between the ages of three and seven, and a typical scenario goes something like this:

A child who lives in, say, Philadelphia, will say to his parents at age four, "You're not my parents. You're Tom and Jane. My parents, my family, are from New York." After listening to this for a year, the parents get disgusted, and say to the child, "Fine, let's go to New York and find your family." They get in a car and drive to New York, and when they get to the area the child says he's from, the child says, "This is it, turn left, turn right. There's my house." The child walks up to the front door, a 42-year-old man answers the door, and the child says, "I'm your wife." When the man says, "No, my first wife died in an auto accident seven years ago, and now I'm married to so-and-so," the child says, "That bitch?" After the child comes in and meets the family, the husband agrees that the child was his prior wife.

This very general example typifies the kind of case that is examined at the University of Virginia, but I don't want to give my reader the impression that there are only 2500 cases like this. There are only 2500 cases like this that the University of Virginia has considered. The fact of the matter is that we have been seeing this kind of event for years. In the 1950s, Carl Sagan suggested that children who claim prior life experiences should be studied. Up until then, these were simply anecdotal events that science couldn't deal with, because there was no way of incorporating a real reincarnation scenario into our science.

In 1958, the American Society for Psychical Research announced a contest for the best essay on paranormal mental phenomena and the relationship to life after death. Dr. Ian Stevenson submitted the winning entry, entitled "The Evidence for Survival from Claimed Memories of Former Incarnations." In that essay, he reviewed 44 previously published cases of individuals who described having memories of previous lives. From the time he wrote that paper in 1960, until he died in 2007, Stevenson worked at the University of Virginia, finding children who claim they had lived before and building a protocol to scientifically evaluate whether there was validity to their claims.

The most impressive cases involved children who were under the age of ten when they first reported the memories; in many cases, these children were age *three* or younger. Dr. Stevenson was surprised by the number of children who, although from very different places, made similar statements about past-life memories. He ended his 1960 essay by saying that the evidence presented did not permit any definite conclusion about reincarnation, but that more extensive study was justified.

One person who read that essay was Chester Carlson, the inventor of the photocopying process that was the basis for the Xerox copier. Carlson contacted Stevenson and offered financial support. This enabled Stevenson to continue his research until 1968, when Carlson died of a heart attack. Since the Division of Personality Studies at the University of Virginia was dependent on Carlson's funding, after the grant money ran out it seemed that Dr. Stevenson would have to return to more conventional research. However, when Carlson's will was read, it was discovered that he had left one million dollars to the university so that Stevenson's work could continue.

That work continues today at the University of Virginia under Dr. Jim Tucker at the Division of Personality Studies. Dr. Tucker also serves as medical director for the Child and Family Psychiatric Clinic.

Typically, the child who reports a past-life memory spontaneously starts speaking about a previous life at two or three years old. Some kids may talk about the life of a dead family member, and may recount details about other family members, events in a previous life, or how they died. These children show a strong emotional involvement with these memories, and often demand to be taken to their previous family. They will persist, whining and crying, until their demands are met. In a large number of cases, parents who take their children to the place they named find that somebody whose life matches details given by the child had died there. When the child meets the family, the child will sometimes recognize family members or friends from that prior life. There can be physical connections as well, such as birthmarks that match wounds on the deceased individual's body.

Ian Stevenson amassed such a large and complete body of information that the prestigious Journal of the American Medical Association said that the evidence was difficult to explain on any grounds other than reincarnation.

And it is not just at the University of Virginia that we see this. A story about a child who remembers being a World War II pilot was covered by both ABC and Fox News. You can find both versions in the video library at www.thescienceofreincarnation.com. Each of the videos will make the hairs on the back of your neck stand on end, in a good way. Let me state this clearly, so there is no misunderstanding. A World War II pilot is shot down and killed. He is reincarnated as a child. That child goes to the Smithsonian Museum, and says "I was in this squadron." The child is brought to the veterans of that squadron at one of their meetings, and recognizes some of them. And those 85-year-old ex-pilots from World War II accept this child as their fallen comrade.

The fact of the matter is, children who claim prior incarnations are a fairly common event. If you go to the video library at www.thescienceofreincarnation.com, you can find several videos of Jim Tucker, Ian Stevenson, and a wide variety of children who claim prior incarnations.

One case study that can be found in Jim Tucker's book, *Life After Life: Children's Memories of Previous Lives* is about John McConnell, a retired New York City police officer, who was shot to death in 1992. Five years after John died, Doreen, his daughter, gave birth to a son named William. William talked about being his own grandfather, talked about being Doreen's daddy and knew things from their shared past only her father could have known. Doreen was ultimately convinced that William was the reincarnation of her father.

The University of Virginia approaches each case with an attitude of wanting to find out and document as much as they can. They are open to all possibilities, including that a paranormal link might exist between the child and the deceased individual, or that it might not.

While I am using one case study from Tucker's book as an example, it is important to note that there are many examples and many subcategories. It is in these case studies and subcategories that we find information consistent with other things we are seeing. I will describe those events in more detail in the coming chapters, but for now, a few examples of what I mean.

Dr. Stevenson wrote a book entitled *Where Reincarnation and Biology Intersect.* What this book specifically covers is birthmarks on children who claim prior-life experiences, and when the cause of death is examined—a shotgun wound, a stabbing—the birthmark corresponds to the death wound of the prior life. That is an example of one subset.

Another subset are those children who I would categorize as cross gender—that is, they remember living lives of the opposite sex. These individuals exhibit cross-gender behavior. Examples include Kloy Matwiset, a boy with a birthmark on the back of his neck that matched an experimental mark made on his grandmother's body. He stated that he wanted to be a girl, sat down to urinate, and repeatedly wore his mother's lipstick, earrings, and dresses.

Another case in that subset is Ma Tin Aung Myo, a Burmese girl who reported memories of the life of a Japanese soldier killed in Burma during World War II. When she was young, she played with boys and liked to pretend she was a soldier.

The reason I want to single out these particular cases is that the coming chapters will show that this type of information has been found worldwide. Not only do we see this effect consistently throughout the world in children reporting past-life experiences, we also see this in people who go through past-life regressions.

The idea that souls can change gender from life to life seems counterintuitive. We shall see, though, that this is consistent throughout all anecdotal reports. The full significance of this becomes apparent when we apply a statistical analysis to this information. The odds against this type of consistent reporting are significantly large. For now, though, let's just stay within this study.

All the events that I have just described, each past life claimed by a child, is an anomaly in science. That is, our current understanding of science cannot explain why we have so many children claiming prior lives, and why it occurs in every culture. When we see anomalies like this, we can categorize it as an effect. We don't know the cause of it. We could say these children have lived before, but we don't know for sure. What we do know is that we are seeing this effect globally in a consistent pattern.

It is not just that we are seeing people claiming to be reincarnated accepted by the family they used to be a part of as that person; we are beginning to understand the science of why we are seeing this.

So it would seem that if you are like the children here described, even if you do not have any memory of living previously, when you die your consciousness will retain coherence. In short, after your death, you will still be aware and remember this life. And now we've gone and put a scientific theory to it, because the science is there in all aspects of this, and it is called the science of reincarnation.

So, in this first chapter, all we have scientifically shown with abundant evidence is that there are a significant number of children who remember having lived before, that this is a global phenomenon, that we do not understand why this is so, and that our current scientific paradigm cannot explain it.

Sometimes in science we see an effect before we know its cause. An example of this occurred in the early 1700s. Farmers would say to scientists, "See this rock in my field? It fell from the sky." Scientists at the French Academy of Science believed the farmers were lying, because they could look at the sky and see that it had no rocks. It wasn't until the 1800s that we learned that those rocks were meteorites that, indeed, fell from space. What the scientists in the 1700s were seeing was an effect for which they didn't know the cause.

The scientists at the University of Virginia are painstakingly aggregating these cases so they can mine the data, looking for patterns, and trying to understand the effects they are seeing. They are using the best scientific methods at their disposal and are leaving a record for future generations of scientists to examine.

This effect that the scientists at the University of Virginia are so painstakingly cataloging neither proves nor disproves reincarnation. This study, however, is but one tile in a larger mosaic. The question now becomes, are we seeing this type of scientific anomaly in other areas? Are there other events or effects that do not fit our current scientific concepts of how reality operates? The answer is that there are several. The next one we will examine is near-death experiences.

Now I highly recommend Jim Tucker's book, as I do everything Ian Stevenson did. It is a small enough amount of the total body of the work done, and it fairly reflects what we are seeing. It should also show the tremendous respect and admiration I have for the work these scientists have done.

If you go to my web site, www.thescienceofreincarnation.com, you can find a video of Jim Tucker talking about the work they are doing at the University of Virginia. I learned from that video that they are loading all 2500 cases into a computer to create a database of 200 variables. At the time of the video they had only loaded 1400 cases into their computers. This is important work. I don't know where this project is at as I write this, but Tucker mentioned how slow and labor intensive it is.

In the foreword, I said that there should be a foundation to fund the science of reincarnation, and this is exactly the type of work that needs to be funded. I told you at the beginning of this chapter how the work at the University of Virginia was initially funded. The people deciding what science to fund should be the people you will read about in this book.

On my web site you can post about this. I invite discussion there. But this isn't the only type of program to fund; this type of study is not isolated. Erlender Haraldsson did a study at the University of Iceland on children who claimed prior lives.

So, if there is all this evidence of children claiming prior incarnations and this is an anomaly in science, something we cannot explain with our current understanding of science, are there similar types of anomalies elsewhere? People who have near-death experiences make reports similar in nature to what we hear from children who claim prior lives. Let's look at that in the next chapter.

CHAPTER 2

NEAR-DEATH EXPERIENCES

If reincarnation is real, then the very mechanism it uses is that our consciousness remains intact after our death. A scientist would say our consciousness is discrete after death; that is, it still retains who we are and our memories of our life. If reincarnation exists, then we are allowed to inhabit another body, albeit with restrictions. It seems as if those restrictions include that we don't remember a previous life. Yet as we have seen, there are those who claim to, children between the ages of two and five. By age seven or eight, these memories begin to fade, so the child can live a normal life free of memory from the prior life.

So in looking at near-death experiences, we are looking for examples of consciousness removed from the body. What happens to our consciousness when we are declared dead, yet continue being aware? In this chapter, we're going to look at the first scientist to really catalogue that experience, and a few instances that were exceptionally well documented.

We all know what a near-death experience is. Somebody dies, is clinically dead, and they see a light; they move toward the light, and they might meet relatives, or meet Jesus or Buddha; then, as they're resuscitated, they are pulled back into this reality. Again, this near-death experience, like children who claim prior life experiences, is an effect that science cannot explain. It is a worldwide phenomenon, and all races and cultures experience it. What I'm going to show you is that all of these events are a mosaic, that they are interrelated, and that there is now a science that can explain their causation.

The book *Life After Life* was first published by Dr. Raymond Moody, M.D., Ph.D., in 1975. It is an investigation of the cases of people who experienced clinical death and were subsequently revived. The subjects come from a variety of backgrounds, but their stories about surviving death share striking similarities.

Life After Life has sold over thirteen million copies in the last twenty-five years. Why is that important? I want my reader to see the interest in and popularity of this topic. In two chapters we will look at another author, Dr. Michael Newton, who also has large sales numbers. Folks, if people weren't interested in these topics, these books wouldn't sell. They are interested because it is a common human experience. Please notice that I am not saying it's true; I'm saying that it is a common effect, something we are seeing that we do not have a scientific explanation for.

Dr. Moody is a world-renowned scholar, lecturer, author, and researcher, and he is widely recognized as the leading authority on near-death experiences. He has taught at the University of Nevada–Las Vegas, where he holds the Bigelow Chair of Consciousness Studies.

Dr. Moody has aggregated anecdotal information from people who either have had near-death experiences or have gone through clinical death and were later revived. What is striking about his research are the remarkable similarities found in the stories of these people. Accordingly, I now provide an overview of these people's experiences, and show the common denominator that Dr. Moody has found in his case studies.

The section of *The Science of Reincarnation* that you are now in is macro observations. That is, these are observations we see in the real world that we can't explain. There are three subdivisions of this section. The first is children who claim prior life incarnations. The second is near-death experiences, and the third is past-life regression, which we'll get to in a later chapter.

Dr. Moody wrote the textbook on near-death experiences, and it should be required reading for anybody who is studying the science of reincarnation. Dr. Moody separated death and near-death cases by the way in which people arrived at them, and they fell into three distinct categories.

First, there are the people who were resuscitated after they were thought to be or were pronounced clinically dead by their doctors. Second, there are the people who, in the course of accidents or severe injury or illness, came very close to physical death. And third, there are people who, as they died, described their experience to other people who were present, people who later reported the content of the death experience to Dr. Moody.

Moody was struck by the number of similarities in these various accounts. In fact, regardless of how a person arrived at a near-death experience, the similarities of the reports were so great that Moody could easily pick out fifteen separate elements which recur again and again in the narratives he collected. On the basis of these points of similarity, we can construct a theoretical experience which embodies all the common elements in the order in which they typically occur. I am about to describe a typical near-death experience. Of the millions of reported near-death experiences that have occurred globally, they all follow the same type of pattern. The International Association for Near Death Studies (IANDS) is an organization for studying and disseminating information on the phenomena of near-death experiences. According to Wikipedia, "A near death experience (NDE) refers to a broad range of personal experiences associated with impending death, encompassing multiple possible sensations including detachment from the body; feelings of levitation; extreme fear; total serenity, security, or warmth; the experience of absolute dissolution; and the presence of a light."

A typical scenario would go something like this: A man is dying, and he reaches a point of great physical distress. Suddenly, he finds himself outside of his own physical body, but still in the immediate environment of where he died. He sees his own body. He could be floating above it, or next to it, and he becomes a spectator to his own death.

He becomes aware that he has "a body," but a very different one from the physical body that he occupied previously, which is now lying dead in front of him. He begins to meet other spirits; they might be relatives and friends who have already died, or it could be a being of light that is warm and kind. This being may ask questions to help him understand what is going on. He can be overwhelmed by intense feelings of joy and love. The being of light may tell him his time hasn't come, and that he has to go back to Earth. He ultimately finds himself reuniting with his physical body.

Common to the descriptions of this is that the light, a nearly universal element, is actually a being that is described as benevolent.

This experience of dying, going outside your body, elevating, seeing and hearing what is going on as doctors try to resuscitate you, meeting the being of light, and then being restored to your body is a common near-death experience. In his book, Moody points out "the difficulty of explaining the similarity of so many accounts. How is it that many people just happen to have come up with the same lie to tell me over a period of eight years?"

Don't you think that there are only a handful of possible answers to this question? Could any of those answers exclude the possibility of something beyond the cessation of physical life?

According to Dr. Moody, there exist realms outside the one in which we live, and the mind will enter these realms upon death. This establishes that the mind—the individual consciousness (as defined before, the soul)—is likely a separate entity from the body; therefore, a physical death does not mean a permanent one. What we can gain from this compelling theory is that if the death of the body does not mean the death of a person's consciousness, then a person's mind can potentially manifest itself in a different body down the road. If death isn't permanent, then reincarnation is possible.

The reports collected from people who have had near-death and clinical death experiences contain a significant amount of common elements. The majority of these reports speak of an out-of-body experience and an encounter with a being of light that, usually, talks to them. The uniformity of the collected experiences, despite the large variance of their sources' circumstances, suggests that the experiences are not merely fictitious or dreamlike hallucinations, but reports of an experience in an actual realm separate from our own, inaccessible to the living, and indicates also that death is not permanent because the mind is able to travel independent of the body.

Establishing the impermanence of death can supplement the theory of reincarnation, for if the mind is able to travel independent of the body, then the possibility exists that it can travel to another body following the cessation of the first. I want to direct you to two good examples of near-death experiences outside of Moody's work.

On April 30th, 1976, Sandra Rogers tried to commit suicide. She describes her near-death experience in her excellent book, *Lessons from the Light*. You can go to my web site, www.thescienceofreincarnation.com, and follow the links to read about her near-death event. What I liked about this particular description was its clarity.

Another one equally as good was the near-death experience of Pam Reynolds. Dr. Michael Sabom is a cardiologist who describes it in his book, *Life and Death*. He includes a detailed medical and scientific analysis of her experience. What I liked about the description of Pam's event found on his website is the pictures of the instruments being used, and how the operation was done.

This operation, nicknamed "standstill" by the doctors who perform it, required that Pam's body temperature be lowered to 60 degrees, her heartbeat and breathing stopped, her brain waves flattened, and the blood drained from her head. In everyday terms, she was put to death. After removing an aneurysm, she was restored to life. During the time that Pam was in standstill, she had a near-death experience. Her remarkably detailed, veridical out-of-body observations during her surgery were later verified as very accurate.

In the 1970s NDEs were simply anomalous events. That is, even though NDEs were becoming more common as medicine got better at bringing people back from death's door, the similar experiences people reported could not be explained. Why do all these people who don't know each other relate a common experience of going through a tunnel and into the light? Why the similarities of the descriptions are so striking and why do the descriptions of the afterlife so neatly parallel the descriptions of the young children in the University of Virginia studies?

So a 501(c)3 foundation was established in 1985 to collect money to fund research into this event, The Near Death Experience Research foundation (www.NDERF.org). In the last chapter I told you about the work being done by Jim Tucker to look at the data for trends and commonality in the cases collected at the University of Virginia. This could and should be funded by a foundation to study reincarnation, but this is all interconnected.

Remember that thus far, the information I am giving you is anecdotal, yet we're seeing similar types of stories coming from different venues: children claiming past lives and people who've had near-death experiences. Serious, long-term investigations are being done at the University of Virginia, and by a doctor at the University of Nevada-Las Vegas. There is, however, a third area to investigate —past-life regressions. Let's go there next.

CHAPTER 3

PAST-LIFE REGRESSION TIME LIVED IN PRIOR INCARNATIONS

While memories of prior lives in children and near-death experiences are spontaneous events, past-life regression is something that's attempted, rather than something that just occurs.

In the first chapter, the lives recalled by children seem to be accepted by a significant portion of their prior families. In near-death experiences, people who have them tend to remember things or know things they could not possibly have known otherwise, except as explained by the near-death experience. However, that is not the case with past-life regression.

> *Past-life regression is a technique that uses hypnosis to recover what practitioners believe are memories of past lives or incarnations, though others regard them as fantasies or delusions. Past-life regression is typically undertaken either in pursuit of a spiritual experience, or in a psychotherapeutic setting. Most advocates loosely adhere to beliefs about reincarnation, though religious traditions that incorporate reincarnation generally do not include the idea of repressed memories of past lives.*
>
> *–Wikipedia*

One of the most credible researchers supporting the efficacy of past-life regression is Dr. Brian Weiss, who has been on Oprah Winfrey's show many times and has published several books regarding some of his subjects and what he's learned from them via regressing them to past lives.

What you hear as Dr. Weiss recounts his sessions is the same thing you hear when children who claim prior lives or people who have near-death experiences report. The descriptions of the environment after death and before life are the same from all the disparate groups. This is a statistical anomaly. Why these reports are similar our current science cannot explain.

Based on what he has written, Weiss is completely convinced that past-life regression reveals the prior lives of his subjects. The overriding question, of course, is whether he is really uncovering accurate information or is he falling victim, as a well-intended researcher, to false results?

Brian L. Weiss, M.D., is a psychiatrist practicing in Miami, Florida. His medical degree is from Yale Medical School, he is chairman emeritus of psychiatry at the Mt. Sinai Medical Center in Miami. Weiss has a private practice in Miami and conducts seminars and workshops, as well as doing training for professionals. He is the author of the past-life-oriented books *Through Time into Healing* and *Same Soul, Many Bodies.* You can visit his web site at www.brianweiss.com.

Dr. Weiss's book is different from references I've previously cited, because instead of giving you case study after case study, Dr. Weiss's book examines one subject and her journey through the past-life regressions for which he was her guide. Instead of multiple subjects with just one paranormal experience each, we get to see one subject examining a multitude of lives that she didn't know existed. Her name is Catherine, and in his book, you can see her discoveries of her past and Dr. Weiss's discoveries about the process.

The seminal moment in Dr. Weiss's psychiatric treatment of Catherine came when he regressed her into her childhood. Catherine had a phobia that Dr. Weiss believed he could solve by hypnotizing her and taking her back to childhood to confront the event that occasioned the phobia. For the longest time, though, her symptoms showed no improvement. Weiss was unable to find an event in Catherine's life that he could link to her phobia. However, he finally had a breakthrough.

His phrase, "Go back to the time from which your symptoms arise," sent Catherine back to one of her previous lives. She answered, and described her prior life. In this way, Weiss regressed Catherine to her prior lives over the course of many sessions. She remembered historical events; spoke in languages that she didn't know, demonstrated talents in other lives she didn't currently possess. It was natural for Weiss to be skeptical. After all, he was a doctor and a scientist. How could he believe what he was hearing? But Weiss kept an open mind and collected more data. Not only about what Catherine was saying to him, but also the other voices that were speaking through Catherine, including one that began to talk directly to him. The voice spoke in such a way that it convinced Weiss what he was hearing was absolutely true; the voice spoke of something only Weiss's closest family members could possibly know.

An interesting note is that during some of the lives Catherine had led, the concept of reincarnation was defined as heresy.

So here a doctor who heads a department at a prestigious university writes about experiences so far outside the scientific mainstream that they seemed like quackery. Yet, courageous people like Dr. Weiss, working in isolated environments, often share unorthodox findings, even though reporting their findings could subject them to ridicule. What began, however, as one individual with the courage to report an unusual result has become a river of researchers with similar findings. But initially, other researchers were unwilling to believe the results.

I mention this because throughout this book you will see a continuing thread—the voices of critics and people who refuse to accept the findings coming from respected researchers and institutions. Later, you will see how the U.S. government is testing this type of ability, while at the same time publicly denying its validity. The government has its own agenda but is impeding research that should be done.

This type of misinformation reinforces misconceptions, and consistent results from replicable experiments continue to be dismissed as wrong. In short, researchers often do not have open minds. This is important to understand because the same dynamic may be at work as you are reading this book. You should ask yourself this question: how can what sounds like fiction be true? If it is true, then the consequences for human awareness are world-changing. You should next ask yourself: if all this isn't true, why are respected institutions and researchers putting their reputations and careers on the line to publish such results? Finally, you should be asking: where's the beef? Where is the hard science to support what I have just read? Rest assured that we will get to that in a moment. First, let's finish with Weiss.

To summarize, *Many Lives, Many Masters* was written by Dr. Brian Weiss, a psychiatrist who found that his patients could recall prior lives when hypnotized, even to the point of being able to speak in a language that they did not currently know. Weiss acknowledges the professional risk he took by writing a book so far outside the scientific mainstream in 1988. But now, that mainstream is changing its flow as reporting on such things is becoming more accepted.

There is one more researcher that we want to visit before we leave this section. If Brian Weiss's area of study is past lives then Michael Newton regresses people to that period of time between lives and gets a similar report of the conditions that exist there. His patients actually describe the levels. I make no claim that these conditions are real. What I say accurately and scientifically that people going through regression report similar conditions whether they were regressed by Weiss or Newton, or had a near death experience or is a child who claims a prior incarnation. The statistical odds against chance make this a very interesting anomaly. So let's look at Newton's body of work.

CHAPTER 4

PAST LIFE REGRESSION TIME LIVED BETWEEN INCARNATIONS

In theological logic, there are three ways in which we know things are true. One is *synthetical*, that is, we can see it through observation—the sun rises in the morning. The second is *analytical*—one plus one equals two. The third is *nonsensical*, that is, we all agree it's true that color is green.

As a scientist, I cannot accept the third criteria, because we can all agree on something and we can all be wrong. We are examining the book *Journey of Souls* because of its lack of scientific scrutiny, coupled with its popularity.

To the scientist, what people are agreeing to—even if they do so without facts, even if they are wrong—can merit study. It allows the scientist to analyze how people think and how they arrive at conclusions. So let's find out what the popular storyline is, because it is useful to know the general consensus and how it lines up with the facts we have.

In this chapter we will look at what millions of people agree is true—that there is a life after death. Which people agree on this? Any Christian, Jew, Muslim, Hindu, Buddhist, or member of a religious group that claims a life after death—in short, a lot of people.

We are going to focus on what happens after you die. *Journey of Souls*, by Michael Newton, presents the case studies of twenty-nine people whom he hypnotized and then regressed to that period of time between their lives on earth. While in deep hypnosis, these people describe what happened to them between their reincarnations on earth.

The previous examples of past-life regression deal with a person being regressed to a past life; for example, the life of a Roman soldier or concubine. In this work, the individuals are regressed to the period between their deaths and rebirths—that period of time when they were in the place called heaven, nirvana, or purgatory.

There are several points to make here. The first is that the mind remains discrete; your consciousness is always you. The second is that the reports—the way people describe their journeys—are remarkably consistent. The third is that their reports are also consistent with those of people who have had near-death experiences. And finally, the data line up with the reports of children who died and claim to have returned in the University of Virginia study.

Michael Newton, Ph.D., holds a doctorate in counseling, is a state-certified master hypnotherapist in California, and is a member of the American Counseling Association. The topic of his book is hypnotherapy for spiritual regression. Newton has also written *Life Between Lives* and *Destiny of Souls*, both of which cover topics related to the book we are discussing. The fact that he has published multiple books on this topic indicates the topic's popularity.

Newton has done past-life regressions the same way Weiss has at the University of Miami, but his area of expertise or interest is that interim period between death and life, when consciousness exists totally in the soul state. In *Journey of Souls,* Newton explains it with a model that describes the mind as "three concentric circles, each smaller than the last." The circles are separated by "layers of connected mind consciousness." The outermost layer is the conscious mind, where our critical and analytical thinking takes place. The second is "the subconscious, where we initially go into hypnosis to tap into the storage area for all the memories that have ever happened to us in this life and former lives." Newton calls the central, innermost circle the "super conscious mind." This is "the highest center of self where we are an expression of a higher power."

First, let me mention the negatives. Every other author I've talked about is connected with a major institution of learning, is totally vetted because other professionals have reviewed his work, and has extensive footnotes at the end of their books so that readers can investigate their source material. When Brian Weiss writes *Many Lives, Many Masters*, he names his patient, informs the reader that he is a graduate of Columbia University and Yale Medical School, and states that he is the chairman of the psychiatric department at Mount Sinai Medical Center in Miami. You can see that his credentials are those of a fully vetted professional, even if you don't agree with some of his assumptions. Not so with Michael Newton.

Newton holds his cards close to his vest. In *Journey of Souls*, he refers to the people as Doctor N or Patient S, making it extremely difficult, if not impossible, to check his facts. I think he should have been more forthcoming; while he might well respond that he wanted to protect his patients' privacy, some patients, if they were simply asked, would have consented to sharing their experiences with others. That's my criticism.

On the other hand, what Newton describes of his patients' experiences is no different than what Weiss describes in *Many Lives, Many Masters*, the experiences of the subjects studied at the University of Virginia, or the information presented by Dr. Raymond Moody. What Newton brings to this discussion is the structure and experiences of what happens between lives.

His books cannot just be dismissed, because if reincarnation does occur, there must be a passage between lives.

It is important to remember that his anecdotal information is consistent with the other, better-documented anecdotal information described in other parts of this book. We can be generous and say that one reason he is holding his cards close to his vest is that critics would rip his patients and destroy his practice. I am sure that if he asked, some of his patients would be willing to be studied, just as the children are being studied at the University of Virginia. In the interest of science, he or someone like him needs to come forward and submit the research to a meaningful peer review.

So, why are we even bothering with this book? First, Newton goes into the afterlife, or the *between* life, in more detail than does anyone else. Second, his detail is consistent with other anecdotal information. It is consistent with the University of Virginia case studies and the descriptions from people who had near-death experiences. It should be noted, however, that the people who have near-death experiences only go a limited distance into the afterlife. But as far as they go, their reports are consistent with Newton's. His reports are also consistent with other past-life regression accounts, and we will soon see how this ties into Weiss's accounts. Finally, his evidence is also consistent with the common elements found across religions; in other words, it is consistent with descriptions of the afterlife reported by religions where judgment or evaluation of a life is reported.

What is important here is not one report, but the fact that all the reports from all sources of anecdotal information are consistent in what they say; they match what the narrative would be if we distilled the cogent points of descriptions of the afterlife from the world's major religions.

Newton has created a landscape of the reality of the soul between lives. The first thing he talks about in his book is how he encountered the research track he's currently involved with. He remarks that what amazed him was the consistency of the reports he was getting from subject to subject. Interestingly, Moody said the same thing about his patients. What I am going to suggest in a later chapter is an analysis of his data, or meta-analysis of all data, using odds against chance as a standard to show whether, statistically, what we are seeing could be true.

All of Newton's test subjects describe a common response. His results are the same results seen with the studies at the University of Virginia of children spontaneously claiming prior lives, or at the University of Miami, where Weiss is getting a similar result. The gold standard in science is replicability from experiment to experiment, and while these are not *experiments* but rather *studies* (because they don't produce an explanation of the effect), we are quantifying this effect throughout a variety of different conditions that cuts across all religions. Whether people believe in God, Buddha, Jesus, or no one at all, they can have the same experience when talking about the soul and reincarnation. Religion, it seems, has a common denominator.

What follows is a brief description of the passage a soul would have from one life to another, according to Newton. He describes it in the following stages.

The journey begins with death in this life, or *departure*. Newton's subjects report a tunnel, as do Moody's subjects. They move to a white light.

Homecoming is the next step. Newton reports that his subjects meet deceased relatives, as do those people who have near-death experiences. A similar experience is also reported by Weiss. The arriving soul is met by a guide.

Orientation is a period of adjustment from a corporeal form to a spiritual one. Newton makes the point that no subterfuge or deception exists in a telepathic world. Souls are not judged, but rather evaluated. At times, Newton talks directly to the returning soul's guide through the person whom he has regressed. Weiss reports the same phenomenon.

Transition is the term used by both Weiss and Newton to describe what was reported to them. Each soul has a *home* with a group of other souls. It is interesting that Newton notes the "commonality" of word usage by different clients to describe spiritual phenomena after they return, they go before soul councils that evaluate a prior life rather than judge it. In many instances, soul groups are described as classes, much like students in a particular class or grade.

Placement is the term Newton uses to describe the soul's level. According to him, there are six levels of souls that indicate advancement, whether you are in the sixth grade or twelfth grade. Each soul has a guide, and the soul's future life selection is based on what that soul needs to work on to develop. There are levels of guides as well, junior guides and senior guides who are graded on how well they help their charges.

Life Selection describes the time when the soul must once again leave the sanctuary of the spirit world for another trip to earth.

Choosing a New Body isn't about just choosing the soul's new gender, but the health and the condition of the body that the soul will occupy. Though we may consider it strange to select a body that is frail or has some physical handicap, it is done to advance the individual soul's learning in the next life. However, gender is a choice. This information is consistent with what is reported by other researchers and other sources.

Preparation for Embarkation involves connecting with the group of souls you will be incarnating with. The issue of how to find soul mates can be complicated. There are relationships based on deep, abiding love, but there are also relationships based upon companionship, friendship, and mutual respect.

Rebirth, we all know what that is.

Since Weiss and Newton worked independently, it seems unlikely that each subject would describe the life between lives in the same way, yet this is exactly what happens. My problem in taking a scientific look at reincarnation is not that just two people report a similar past life regression, but that so many are reporting this scenario; if I tried to cite all of these cases, my book would never get past all the common stories.

I am not asking you to believe any of this. I am writing a book on science, on evidence. I want to emphasize that this is a common human response of subjects undergoing past-life regression.

So I now want to direct you to the similarities between Weiss's results in regressing Catherine and Newton's results. The information coming from Catherine cited the same idea of levels of development, the same idea of choosing how you return, the same ideas of learning and growing, a repeated process. The stories that Weiss and Newton report regarding the place between lives are consistent with each other.

CHAPTER 5

CONSENSUS OBSERVATIONS THE ELEPHANT IN THE ROOM

I could've done the summation at the end of the last chapter, but it is really its own point. So let's recap.

Chapter Number 1—children who claim prior lives are just the most scientifically aggregated sample of a field that is filled with them. It is common lore. We all know, or we know someone who knows someone, who had this happen to them. The fact that it was so over-reported is what led to the studies at the University of Virginia to begin with. It is an anomaly. We cannot explain it with our current science.

Chapter Number 2—near-death experiences are a global phenomenon, and they share a common narrative. We're trying to study it with various clocks in the room, testing certain meters to see if we can detect an electromagnetic presence, but it is common experience. People who experience this all describe it in the same fundamental way.

Briefly, I want to mention that some scientists claim the tunnel and light we see at death is actually caused by the brain no longer receiving oxygen, and chemicals in the body. The issue then becomes why people who are regressed have similar experiences? Those people are not dying. The point I want to make is that similarities from all these experiences must be contrasted and compared, and dismissing one effect we see is poor science when that effect repeatedly presents itself in a similar way in other macro-observation events.

I am not discounting the science these scientists used, but in order to have a truly reliable scientific view then all observations must be examined, not just some of them.

Chapter Number3—past-life regression is a common experience, yet again, it is an anomaly, because we haven't proved an afterlife. But people regressed can go back to multiple prior lives.

Chapter Number 4—regression to between lives is also common for people who undergo past-life regression. These people not only tell common stories but also use similar words in expressing a common location or activity. Now here is the elephant in the room. If we look at descriptions from all three of these experiences we find an overwhelming amount of similarities in the stories these people tell. In the coming chapters we are going to look at this statically and figure the odds against

CONSENSUS OBSERVATIONS.... 37

chance this would happen. While that in and of itself is only one indicator, we are going to compare that to the odds against chance in the coming section, extra sensory perception.

You can go to the film library at the web site www.thescienceofreincarnation.com and see videos of the scientists I have mentioned so far describe what they have found, as well as hear first-person accounts of individuals who have had these experiences.

In a few chapters we are going to get into the statistical analysis scientists use to determine probability. The odds against chance of this being a coincidence—that is, that the similarities of these stories being just chance or an accident—can be measured. One of the points this book makes is that this information should be studied as a whole, as opposed to the way it is viewed now, anecdotally and separately. Let me show you how this works in just one category, past-life regression.

Past-life regression has the most interesting example. First, you have Dr. Weiss at the University of Miami regressing people to remember not only prior lives, but also introduces us to beings who manage the period of time between our death and our rebirth. Kudos to Dr. Weiss for accurately reporting his findings, standing up for it, and having the courage to clearly report what he is seeing, which has been validated by other studies.

Dr. Newton, not so many kudos to him, because he doesn't allow peer review, even though he is regressing people. However, he is regressing people to the period between their death and their rebirth, and that is exceptionally interesting—not for the research he produces, but for how it fits so neatly into the other pieces.

If we are putting together a jigsaw puzzle of the whole of the science of reincarnation, then this piece of observational data is invaluable.

I strongly recommend that Dr. Newton open up his research to others who can study his methods; additionally, talk to those individuals who are willing to come forward, because there are a multitude of them.

Is it possible to code these cases for the same 200 variables that the University of Virginia is using? If the statistical results point to the odds against chance being high, can we reasonably expect that this is true?

That all notwithstanding, the point of this short chapter is to show how important it is to focus on the whole, rather than the individual pieces of the mosaic. What we have are children who couldn't possibly know anything about events, because they're only three or four, yet they are able to tell the researchers what is going on in the family dynamic more accurately than the researchers can know. People who have near-death experiences passing away, meeting relatives, and returning, so there is an image there that they see, the same as the children in the other study, and people who are regressed. There is a landscape of the world between death and rebirth that is being reported *consistently* and in the same way across all of the studies.

So the question is, first, if this is the human condition, why if we are aware here, could we not be aware there; and second, can we explain it with our current understanding of science?

The answer is that a theory, a real scientific theory, is emerging based on lab results whose implications are beyond huge. It would be a rebirth of awareness of our place, not only in the universe, but in time as well.

In the next chapter, I am going to start showing you how this theory began to evolve as other aspects of human consciousness came to be proven. That emerging science can support this existing observation base, because it allows for the idea that our mind can exceed and reach beyond our bodies, and we have now quantified its effects well enough to be satisfied with the results. This is beyond a mind fuck. It's a mind birth.

So how do you explain, using our current scientific knowledge, what we are seeing as a common and global effect? Everyone, every culture, every spot on the earth, sees these events. It is not about being a certain religion or race or nationality. It is about being human. This affects us all. It treats us all the same. It reduces us all to one common denominator. If we are all different denominations of religion this at its core reduces us to one common denomination.

The common denomination is simply being human. The only assumptions it can have are those that apply to us all. One of those assumptions is that we all have been given freedom of choice. Some of us may be in situations where we are not able to express that freedom, but inside, where no one can see, we decide. Every reader of this book will ultimately be able to make his own choice if this is true.

One last comment before I leave this chapter. Doing the research and writing caused me to pause and evaluate my life on the basis of the life-review committee. It made me, not depressed, but pensive. The way I would describe that to you is similar to what people who experience near-death experiences say about a change of attitude about life. What I am telling you is personal about me.

This information caused a change in my attitude about life. Some of that change was uncomfortable to me personally.

There were people who I never want to meet again, but apparently they are still going to be around. And while it is not a judgment panel per se, I will have to account for my actions.

Now if every macro observation carries the same type of description, then the point is not that I believe it. I am a scientist, and it gives me pause, and I look to probabilities and statistics.

If part of what I do here effects what happens over there, haven't I just described the interaction of two subatomic particles?

It causes me to look for a structure that might support those observations, because that is how science is done. I keep coming back to this folks; this is the science of reincarnation. A science is made up of disciplines; one discipline in the science of reincarnation is past-life regression.

Dr. Newton and Dr. Weiss are both practitioners of this specific discipline. Dr. Weiss does prior incarnations and Dr. Newton specializes in that time between incarnations. Dr. Newton does not want you to see who his subjects are; he admirably protects their privacy. Weiss on the other hands submits his results for peer review. What is compelling is the similar nature of what their subjects say, even though these two researchers have never worked together. How then are we to scientifically examine their findings and analyze the results?

While I disagree with Newton on limiting exposure to his source material, he provides a very insightful and well-followed experience of that time between lives.

What I suggest is that he collates his information with Dr. Weiss, who is a continent away, to look for similarities, and to devise a better interview system using the best features of both of their systems, and sharing data. Because as of now, nowhere is there a model of what is happening after we die based on their collective findings.

Here's what I want you to get from this section. First, real scientists at major universities are putting their reputations and precious funding on the line to research projects they think are real, despite being out of the mainstream. These are serious researchers who are trying to understand events that they see. Second, the institutions they work for—the University of Nevada, the University of Virginia, the University of Miami, and Sinai Medical Center—take their research seriously. Third, their findings are no longer being viewed as paranormal or quackery, for all their findings are interrelated.

Another thing to note: just like the similarities in the stories involving the children who had prior-life experiences, or the people who had near-death experiences, there is a remarkable similarity in the way people describe the afterlife. In Dr. Tucker's book, Life *Before Life,* William, whom we met in Chapter 1, says, "When you die, you don't go right to heaven. You go to different levels—here, then here, then here."

Now, here, in *Many Lives, Many Masters,* is one of the voices that spoke to Weiss through Catherine: "There are seven planes . . . seven through which we must pass before we are returned."

Then in the next chapter Newton's patients actually name the planes or levels. The odds against this type of statistical anomaly are quite high.

So to conclude this section on macro observations, we see three separate effects of events that lead us to similar descriptions of what occurs after death. By having an accredited science of reincarnation we could begin to see the whole picture of what occurs to us all globally every day. We cannot explain this overall effect we are seeing with our current science. But because we do not have a science of reincarnation, we cannot study this effectively. We also cannot incorporate other events we see in other sciences.

If we reincarnate, then we should be able, with only our minds, to reach outside our bodies. A question to ask at this point is whether there are other anomalies in science where, instead of the event happening to us, we create and control (at least to some extent) the event? In fact, by the early 1970s this type of anomaly was being seen in the science of archeology.

CHAPTER 6

CLAIRVOYANT ARCHEOLOGY SECTION 2 EXTRA SENSORY PERCEPTION

It is this chapter that begins to turn all this information from just anecdotal reports to true analytical science. But before I start, I want to thank Stephan Schwartz for his permission to quote liberally from his book *The Secret Vaults of Time*. His book was pivotal in my writing *The Science of Reincarnation*. His daily E-report, the Schwartz Report, at www.Schwartzreport.net, is daily reading for me. I cannot recommend this man highly enough. The following is excerpted from *The Vaults of Time*.

"It is my conviction that I have received knowledge about archeological artifacts and archeological sites from a psychic informant who relates this information to me without any evidence of the conscious use of reasoning. . . . By means of the intuitive and parapsychological, a whole new vista of man and his past stands ready to be grasped. As an anthropologist and as an archeologist trained in these fields, it makes sense to me to seize the opportunity to pursue and study the data that was provided. This should take first priority" said Professor Emerson.

In 1973, J. Norman Emerson, a senior archeologist and professor, and perhaps the most respected one in Canada, endorsed using psychics in archeology. This break from orthodoxy, advanced by a man in the last years of an influential career, was groundbreaking.

"I have to confess that after 30 years of work, and, let's face it, as one of the few real experts in my special field of Ontario Iroquois Indians . . . well, prior to 1971, if you had asked me, I'd have to say that of those questions that there was no way I could have even have attempted to answer them. Today, however, I would reply that yes, it may be possible to do so—with the help of psychic persons."

Traditional research into prehistory, he would later say in explaining his motives, is pretty solid. We are able to trace with a great deal of confidence the story of what happened and at what time in the history of man such events took place, as well as how extensive such happenings and developments were over a geographical range in territory. But traditional research into human prehistory still has a major weakness. There is a lack of humanity. Real men and women hardly ever tread across the pages of our site reports, conference papers, and written books.

"We move from questions of when, how widely, and what happened in the past—where we have some confidence in our findings—to questions of what did it all mean and of what value it was, with less and less assurance and greater speculation. As the realms of art, symbolism, social meanings, and individual and societal values are encountered, our ability and confidence vanishes. Yet, these are all questions which make a huge difference when one tries to understand the living person and his culture."

J. Norman Emerson, senior professor of anthropology at the University of Toronto, founding vice president and former president of the Canadian Archeological Association is considered by many the father of Canadian archeology.

Close to 90 percent of that nation's professionals in the fields of anthropology and archaeology had trained with Emerson at some point in their careers. His former students include other University of Toronto anthropology faculty members, directors of the National Museum of Canada, professors at other universities, and most government archeologists. His impact on Canada's research in those disciplines has been immense, so when in March of 1973 at the annual gathering of Canada's Archeological Establishment, when he stood to speak, some wondered why a scientist would in the last years of an influential career, suddenly dessert 30 years of orthodoxy for something like a psychic. His answer was because it was the truth."

I am not going to recount the history of psychic archeology here, but I will tell you that *The Secret Vaults of Time* is a wonderful book, and that the stories it brings to you are true. Emerson's experience with the paranormal began in the 1960s, but it wasn't until he met George McMullen that things that would shake the foundation of archeology began to change.

I have a lot of ground to cover in this book, so I'm not going to tell you how Emerson and McMullen came together. Instead, I'm going to jump to how they worked together. Emerson would give McMullen an artifact, like a clay pipe or the shard of a bowl. Such artifacts are known as lithics. McMullen would turn it over in his hand, and he would tell Emerson how it was made, how it was used, what Indian group used it, where it was found.

As Emerson said when delivering his paper, McMullen's accuracy ran to about 80 percent. But one day, a man by the name of Jack Miller came to Emerson and asked if McMullen could solve a little mystery. Miller brought with him something made of black stone, about two-thirds the size of an average man's palm, and flat on both sides. It was generally agreed between Miller and Emerson that the stone was argillite. Argillite is a shale-like rock found on the Queen Charlotte Islands off the coast of British Columbia; it is about the hardness of soapstone, which makes it a good material for carving. Miller told Emerson that he knew where the stone had been found—at the bottom of a posthole he had excavated at a site near the town of Skidegate on the Queen Charlotte Islands—and the time period in which it had been worked. Later that evening, Miller gave the piece to McMullen. "The stone, McMullen stated, with the certitude possible only to one totally ignorant of the intellectual information on a subject, was carved by a Black man from Port-au-Prince in the Caribbean, from whence he had been brought to Canada as a slave."

Emerson was appalled, as he later admitted. Here I had just presented a paper on how good George was and how you could use psychic data, and just a short while later, he was saying something that was patently ridiculous."

Here, then, you have a psychic providing information that flies in the face of established orthodoxy and historical information. There were no black men, historically, on Port Charlotte Island in that time period. McMullen wasn't just wrong. He was completely wrong. It turned out, however, that it wasn't McMullen who was wrong. It was the archeologists who were wrong, and that would be proven.

The first thing that happened was that this lithic was given to other psychics. He gave it to a woman who knew nothing of McMullen's response, and "to his astonishment and excitement, she began immediately telling him that it had been carved by a Black man from Africa who had been the victim of a vast sweeping slave trade, and had been brought to the New World. As she went on, she gave many details about incidents McMullen had not mentioned, but she also corroborated many of the features that he had described."

Other psychics began to confirm McMullen's analysis without ever having spoken to McMullen. This confirmation, however, wasn't the proof.

"Almost two years later, a team of physical anthropologists totally unconnected with Emerson went to British Columbia to do blood analysis on the Indians of the area. Their report, when filed, contained what was for them the disturbing observation that in an area where no Blacks were supposed to have been until Modern Times, one tribe showed unmistakable evidence of a Black forbearer. The tribe in question was the one into which Emerson's psychic team had said the escaped African had married."

This was physical DNA corroboration that a black man had lived with that tribe at that time.

For any readers who are not familiar with DNA and blood markings, you can actually see in the DNA of Asians the marker of Genghis Kahn. Because he fathered so many people, you can see a similar DNA marker among descendants from that period of time. Let me make it clear: the DNA records of the Iroquois Indians contain proof a black man lived in that tribe during the period of time in question; it's irrefutable.

To conclude, we have a respected archeologist using a psychic who, when given a lithic, comes up with a reading that is totally outside of what could possibly be accepted. But it turns out that not only is that same information confirmed by other psychics, it's confirmed by an independent, unrelated research team doing blood analysis of the Indian tribes.

You then have a man who was at the time the most respected archeologist in Canada stating unequivocally that this is the best method to use to study archeology, and he single-handedly ushers in a new era in archeology, which is psychic or clairvoyant archeology.

If you don't believe I've just proven clairvoyance, read Schwartz's book and come back to mine. I need to cover ground. He makes the case convincingly, but read him if you don't believe me.

But now let me tell you what this has to do with reincarnation. Several years ago, *Discover* magazine ran an article about the twenty most important science books ever written. They included *Origin of Species* by Darwin, *Relativity* by Einstein, and *Principia* by Newton. On the list there was a book entitled *The Structure of Scientific Revolutions* by Thomas Kuhn. I am a science buff, but I had never heard of it, and I was amazed to find out it was written in my lifetime, in 1962. Of course, I immediately bought it and found it to be a slow, difficult read. What I could not deny, though, was the brilliance of the work, even if I initially struggled with it. It went on my bookshelf and stayed there.

Later that year I was reading another book, *The Secret Vaults of Time* by Steven Schwartz. It is a history of clairvoyant archeology and a fun, easy read.

Clairvoyant archeology is one method currently used by archeologists to determine where to put the shovel in the ground. Above you read about the beginnings of clairvoyant archeology; now a standard method, it works like this: three clairvoyants independently look over maps of the archeological site to identify hot spots—where the archeologist should dig to find what he is looking for. Then the maps are overlaid with each other to find common hot spots. This produces a consensus map, if you will. A clairvoyant is then brought to the site at the start of the dig to locate the single best place or places. *The Secret Vaults of Time* traces the evolution of this method from Blair Bond in 1900 in Glastonbury, England, to Alexandria, Egypt, in 1995.

But at the end of his book, Schwartz does something remarkable. He references Kuhn's book to show how this now accepted method of archeology is a scientific anomaly, in that the method lies outside the standard paradigm of science.

According to Kuhn, no science becomes a science until it has a paradigm. This includes anthropology, archeology, biology, chemistry, geology, hematology, psychology, physics or any other science you can name. Most research is done to support the accepted paradigm—to explain it and expand it along the lines that the paradigm predicts. But some results require that the paradigm change because experimental results contradict it.

On rare occasions, the *metaparadigm*—the paradigm that overrides all others—is forced to change. This change can take centuries or occur very quickly. Such revolutions are rare and far reaching, and we are on the precipice of such a change. So what is a metaparadigm? There have been two main ones in three thousand years. Let me explain the great magnitude of the change we are on the verge of.

The Genesis metaparadigm is the first one. From the words "God created the heavens and the earth," until the middle of the 1600s, all scientific research was done under this premise. Copernicus was afraid to publish his results because they contradicted this metaparadigm. His works were not published until after his death. Galileo was put under house arrest for publishing results that contradicted the Genesis metaparadigm. Most scientists of the seventeenth century accepted the calculation of James Ussher, an Irish bishop, who added up all the generations of the Old Testament and determined that the world was created in 4004 B.C. Ussher's work is an example of something that supported the Genesis metaparadigm.

The *grand material metaparadigm*, the one we work under now, is the ruling metaparadigm. Schwartz explains that "it has at least four critical assumptions. They are: 1) The mind is the result of physiological processes governed by bioelectrical postulates; 2) each consciousness is a discreet entity; 3) organic evolution moves toward no specific goal but simply flows according to Darwinian survivalism; and 4) there is only one space-time continuum and it provides for only one reality."

His point is that the current metaparadigm cannot be reconciled with the anomaly of using clairvoyants in archeology. Our current scientific theory has no place to incorporate a human mind reaching through time and space and extracting accurate information. Previously, the results of clairvoyants could not be tested scientifically, but this changed with their use in archeology. He explains: "This was testable on four different levels: (1) testable information on site location;

(2) testable information on surface geography and subsurface geology; (3) testable information on what artifacts and remains will be found and at what depth; (4) testable positioning on bone finds; e.g. the broken blue edge points straight down."

When a clairvoyant says dig here and points to a spot, and the archeologist then digs there and finds what he is looking for, it proves the method works. When a clairvoyant interprets what the archeologist finds, contradicts the current interpretation, but is subsequently proven right, it validates the method. Clairvoyance has been accurate enough to be used in a standard way by many archeologists and to be accepted as a standard practice by the archeological community.

This change occurred in 1974 and was lead by J. Norman Emerson, as described above, who at the time was the president of the Canadian Archeological Association and is considered by many the father of Canadian archeology. For the first time, we could clinically test clairvoyant knowledge. It first met with resistance by the archeological establishment, but thirty years later it is an accepted methodology.

But let's stay in 1974 for now. If the mind could reach out and receive information, could it also send information? Five years later researchers at Princeton University designed an experiment that altered the way we perceive this process.

At the beginning of the next chapter, I am going to say that clairvoyant archeology proved a person's mind can reach through time and space. It'll be in the very first sentence. Some might feel that prove is a strong word, but prove is the word I use and I ask you to judge for yourself if I am right, based on four criteria.

First, Emerson himself felt it was proved and said as much. So whatever he felt about this, I agree with.

The second criterion is not to see this argument as a series of steps but rather as one fabric. Some explanations for events stretch our understanding of reality, but ask if what we see fits with everything else we see. In short, does it hang together as a theory? Occasionally, extraneous information from other sciences explains what we see; in short, support in the microscopic world explains what we see in the macroscopic word.

The third is odds against chance: what are the odds against chance that this information would line up in this way?

Fourth is the results of the Princeton studies themselves, which verified in the lab what the archeologists were seeing in the field.

So, based on my own examination of the above criteria, I chose the word "proved." Before this book is finished I will examine the remaining criteria and you can judge for yourself.

CHAPTER 7

THE INTENTION EXPERIMENTS SECTION 2 MICRO OBSERVATIONS

So the question was simple. If we could receive information, could we send it? We still didn't know where it came from or went to—the zero point field, dark matter, probability? We didn't know. But we found out conclusively each and every one of us can send information.

So if we could receive information as demonstrated in the last chapter, that is, if clairvoyant archeology proved that a person's mind is able to reach through time and space, the next question was, could we send information in the same way? How could we test this ability in the lab? For an answer, let's see what began to happen at Princeton University beginning in 1979.

The Intention Experiments began at Princeton University and were lead by Robert Jahn and Brenda Dunn. If you go to their web site and watch the fifteen-minute video, you will hear Brenda Dunn say "This typifies the dramatic results they have gotten, results that have yet to be integrated into the minds of the current generation of scientists."

I am going to explain the results in layman's terms and I am going to be purposefully vague, and I make no apologies for this, because this is not about my being precise, it is about explaining the overall concept with generalities so anyone can grasp what is going on here, whether they have a scientific background or not.

These experiments involved programming a computer to produce randomly an equal number of zeros and ones every hour. Every hour there would be 50 percent zeros and 50 percent ones. The computer was then connected to a screen that would show two different pictures, for example, a tree and a boat.

This is what the researchers' subjects, average people recruited off the street, would see when they interfaced with the computer. People would sit in front of the screen, and Jahn and Dunne would ask them to make one picture appear more than the other by their intention alone. People could close their eyes and think "tree, tree, tree," or they could talk to the computer out loud. They were not allowed to touch the computer, so the only way they could affect the computer was by their intent—their thoughts.

Here are the results. Virtually everybody could make one picture appear more than the other by a margin of 52 percent to 48 percent. If a bonded couple, a man and a woman, sat in front of the computer and did the experiment jointly, the researchers found the computer would produce 54 percent of one picture and 46 percent of the other. If two women, however, sat down together to attempt the experiment, they would get the 54/46 result, but sometimes in the wrong direction.

Couples would influence the machines six times as strongly as individuals. Also, if a couple was not in a relationship, they would still have a complimentary effect on each other. Men had a better chance of getting the machine to do what they wanted, but women had a stronger effect on the machine, though not always in the direction they intended.

This was not an isolated study. These experiments started in 1979 and ran until 1994. Other labs running the same type of experiments got similar results.

The numbers of the total data are interesting. It was a twelve-year period, nearly 2.5 million trials. Of all those trials, it turned out that 52 percent of the trials were in the direction that was intended.

By intention alone, participants in the experiment were able to bend the computers, at least a bit, to their will. They had some influence on it a significant percentage of times. These results, by the way, have been submitted to the U.S. National Research Council, and the council has concluded that the trials Jahn and Dunne conducted could not be explained by chance.

So if it's not chance, what is it? How, on a scientific level, can you explain the effect of influencing a computer by simply wishing?

On the Science of Reincarnation website under Government Studies and Video Library, you're going to find a link to the Princeton University PEAR lab; there you can watch Robert Jahn and Brenda Dunne explain what it is they do.

The best explanation offered on how Jahn and Dunne arrived at the results they got is as follows: Subatomic entities can behave either as particles or as waves. A particle is a precise thing with a set location in space. A wave is diffuse and unbounded, and has a region of influence, which can flow through and interfere with other waves. Jahn and Dunne feel that consciousness has a similar duality. Each individual has its own particulate separateness. That is, you are a defined thing in space, but you are also capable of wavelike behavior, which could flow through any barriers or distance to exchange information, and interact with the world.

At certain moments, this wavelike consciousness can get in resonance with, that is, have the same frequency as, other subatomic matter.

What Jahn and Dunne seem to be saying is that you and the computer develop coherence. That is, the wavelike component of your being gets in resonance, and one can influence the other.

The question we should ask is: can we believe the results they got? Answer: absolutely! They used over a quarter of a million subjects over a twenty-five-year period, and after publishing their results, other labs tried to duplicate their experiments and got the same results. So, we are seeing an effect that we don't quite understand.

Dean Radin said in his book, *The Conscious Universe* (which we'll discuss later), that

> Just as a photon is both a particle and a wave, perhaps consciousness too has complementary states. In ordinary states, the mind is more particlelike and is firmly localized in space and time. This is supported by the ordinary subjective experience of being an isolated, independent creature. But in unusual, nonordinary states of awareness, our minds may be more wavelike and no longer localized in space or time. This is supported by subjective experiences of timelessness, mystical unity, and psi.
>
> As with particle-wave duality, it is not the case that only one or the other description is true, but *both are true at the same time*. The fact that we have trouble thinking in terms of "both" rather than "either-or" says more about the limitations of language than it does about the nature of reality. If our minds have complementary

characteristics, then perhaps we can be more particlelike or more wavelike depending on what we wish to be, or what it is suitable to be at the time, or what we are motivated to become."

If we can influence a machine, can we influence another person or a disease? In fact, we try it all the time. You have all heard about the power of prayer; now it's being taken into the lab and dissected.

In each of the preceding chapters, I have selected one or two examples of each category. Whether it involved research at the University of Virginia, past-life regression, or near-death experiences, I have tried to give a flavor of the research that's being done in each category, without overemphasizing the technical nature of the science.

I just showed you how Jahn and Dunne documented that subjects were able to influence a computer, but could that influence extend to biological entities?

Let's pause for a second to talk about Lynne McTaggert. I learned about the following experiments from Lynne McTaggert's book *The Field*. I recommend everybody pick up a copy of *The Field*. It is simply excellent.

There is one important point here that should not go unrecognized. If Jahn and Dunne first found and proved this effect, it has been Lynne Mctaggert working with it. Some day her results should undergo a meta-analysis along with Jahn and Dunne's results. It is combining information and studies like this that will allow us to fully understand what we are seeing and be sure of our assumptions and results.

The information McTaggert discusses in terms of these experiments are available from a variety of sources. You can find this information both on the web and on my web site, under Studies.

Many experiments done over the last 20 years—most notably Jacques Benveniste's work at INSERM, The French National Institute of Health and Medical Research—has been written about in magazines and journals, so there's abundant sourcing for original material, but Lynne aggregates it in a particularly insightful way. My guess is that by the time this goes to print, there will be a link from my website to hers, because she is on the forefront of using what Jahn and Dunne discovered in the Intention Experiments.

Intention studies did not start at Princeton University. Princeton University started their studies in 1979. A decade earlier, Doctor Bernard Grad of McGill University in Montreal was interested in determining whether psychic healers could actually transmit energy to patients. Grad had slowed down the growth of plants, by soaking their seeds in salty water. Before he soaked the seeds, he had a healer lay hands on one container of salt water, which was used for one batch of seeds, and soaked the other batch in salt water the healer hadn't touched. The batch exposed to the water treated by the healer grew taller than the other batch.

Grad then did a study to see if the process worked in reverse: would negative feelings have a negative effect on the plants? A man who was being treated for psychotic depression laid his hands on the water for one batch of seeds. Those seeds grew less than those that had been soaked in water that had gone untouched by the depressed patient.

Grad analyzed the water with infrared spectroscopy. The water treated by the healer displayed minor shifts in its molecular structure. It also had decreased hydrogen bonding between molecules, which are similar to water, that has been exposed to magnets. All of this has been confirmed by other scientists.

Grad also found that healers could affect results by intent alone. After controlling for a number of factors he showed that mice with skin wounds healed more quickly when healers treated them. He also found that healers reduced the growth of cancerous tumors in laboratory animals.

So, if we can affect plants and rodents simply by intention, the question becomes: can we affect humans as well? The scientific tests that followed led to an amazing conclusion. The two examples I give you now are typical of results other labs have gotten. These are not isolated findings.

In 1988, a doctor named Randolph Byrd ran a randomized, double-blind study to investigate remote prayer. About half of almost 400 patients in the coronary care unit of a hospital were prayed for by a group outside of the hospital. When the experiment commenced, there were no significant statistical differences between the group that was prayed for and the control group, which was not. The prayer went on for ten months, at the end of which, the group that had been prayed for were doing significantly better than the other group: their symptoms were less severe; they required fewer treatments with antibiotics; spent less time on a ventilator; and had fewer cases of pneumonia.

When you pray for something to happen, you intend or wish for it to happen, and whether you're praying to Jesus, Buddha, or an ancient rain god or some other diety, you're still positively affecting the object of your prayers. What was different in this case was that prayer was being taken into the laboratory and tested.

Studies like this were a preamble for Elisabeth Targ's and Fred Sicher's study, who brought together a variety of healers: Buddhists, Christian healers, evangelicals, a Jewish kabbalist, a Sioux shaman, people who worked with crystals and bells, and a practitioner of Qigong. The only absolute requirement was that each healer must truly believe that their method worked.

All healing was done remotely, and for subjects, they used a group of advanced AIDS patients. The target and control groups were matched as much as possible in symptoms, T-cell counts, degree of the illness, etc.

When the six-month study was over, 40 percent of the control group was dead, but all ten in the group that had been the target of the remote healing were healthier than they had been at the beginning of the experiment.

No matter what type of healing they used, no matter what their view of a higher being, the healers were dramatically contributing to the physical and psychological well-being of their patients. Another study, by the Mid-American Heart Institute, yielded similar results.

If you want to know more about these studies, please read Lynne McTaggart's book, *The Field*, because she aggregated this type of information well. For the purposes of this book, though, one conclusion is inescapable. We began by showing that we can receive information through clairvoyants, as archeologists do. Next, we saw that we could influence a computer merely with our thoughts. Then we saw how Grad demonstrated that we can influence the heath of plants. What occurred in the preceding two studies is that we have factually proven that prayer works. This is not something we need to take on faith any longer. It has been taken into the lab, quantified, and shown to work. The experiments meet the gold standard of scientific research, which is repeatability. It can now be accepted as fact that prayer works. A theory regarding how it works is now coming slowly into focus.

We have already shown that you can influence a computer. The theory is given that your consciousness is both a particle state, you, and a wave state at the same time. You are able to get in coherence with a computer to effect change. In prayer a bunch of people gather all coming together on their own wave length so to speak. In praying together they sync their wave states by collectively focusing and thereby are able to effect change as a group. They literally get on the same wavelength.

There is one last note that I would like to provide on this from McTaggert's book. "Studies of the nature of the healing energy of the Chinese Qigong masters have provided evidence of the presence of photon emission and electromagnetic fields during healing sessions. These sudden surges of energy may be physical evidence of the healer's greater coherence—his ability to martial his own quantum energy and transfer it to a less organized recipient."

The standard method of operation of healers actually suggested that individual consciousness doesn't die. One of the first serious studies of mediums, at the University of Arizona, suggests that consciousness continues after death. Even after researchers took thorough steps to prevent any kind of cheating, the mediums in the study came up with roughly eighty different details about the deceased people they were trying to contact. Relatives of the deceased confirmed accuracy of about 83 percent of what the mediums reported. Accuracy rate for non-mediums used as a control group was only about 36 percent. In face of these laboratory studies, it is hard to come up with any conclusion other than that the mediums were communicating with the dead.

In McTaggert's book, Fritz-Albert Popp describes death as a "decoupling of our frequency from the matter of our cells."

Pam Reynolds, whose near-death experience I discussed in Chapter 2, said that going back into her own body was like "jumping into a pool of ice." Here, you have a scientist describing it as a decoupling and Pam describing the same separation in a different way.

This book is about the *science* of reincarnation. The ability of the human mind to reach across time and space, an ability we all possess, is now indisputable.

At this point, let's define the *soul* in very general, easy-to-understand terms. The soul is simply your mind disembodied. A soul is what leaves your body, but it is your own consciousness. Scientists would call it a discrete consciousness, but that simply means that you retain your personality, your memories, and your frame of reference. Your soul is you without your body. So, if your mind and soul are synonymous in that way, then where is your mind located?

If your mind is defined as your consciousness, will, and memories, then you need to isolate its components. Let's take memory. The first reaction of 99.5 percent of readers will be that memory is in your brain. As we'll find out, that's not necessarily so.

We first need to see the way the quantum world of the soul—of wave forms, frequency, brain waves, or whatever term makes sense to you—interacts with the macro world of people. But before we attempt this, we need to survey the landscape the soul would navigate.

In our *particulate form*—our bodies—we see reality. Look around—what you see is the body's reality. But if the soul exists, it exists in a different reality. Science is peeling away that reality like an onion, one layer at a time. If our minds, our souls, exist without our bodies in a wave form, we need to understand their *landscape*—their *reality*—before we get to how we, as souls, move from one reality to another.

One last thought. We have scientific evidence that prayer works. It is taken from the laboratory, it is replicable, and it works the same for everyone. So any religion that tells you to pray and that prayer works can now refer to this science and say, see, here is the proof.

But how certain can we be that this human ability to reach outside our bodies is true?

CHAPTER 8

PSI
HOW CERTAIN CAN WE BE THAT THIS INFORMATION IS TRUE

What we have seen in the last two chapters are examples of what scientists call psi. It is the demonstrated ability of a human mind to reach through time and space. There were two examples, one with clairvoyants, and the other with the work at Princeton University. In the Princeton University chapter I began to touch on the scientific theory that is beginning to evolve around these abilities. I will get into that more later, but I want to do a few things before we proceed.

I want to give you a bit of a background about how long this type of research has been going on, and I want to give you a sense of how reliable this information is. In short, I want to show you that you can rely upon the fact that this human ability called psi is true and accurate.

What does this have to do with reincarnation? Reincarnation requires our mind to retain coherence—that is, continue to be us after we die—so that we can then come back and inhabit another body. If our mind can extend outside of our body in life, then why can't it do it in death? You can say that psi does not prove our ability to have a consciousness after death, but if it does, it would make sense that it would possess that ability in life. So let's see if the scientific community accepts that ability, and see if you agree with them.

The book that explains psi research as well as any I have read is Dean Radin's *The Conscious Universe: The Scientific Truth of Psychic Phenomena*. Radin has a master's degree in electrical engineering and a doctorate in psychology from the University of Illinois. He has worked at AT&T's famed Bell Laboratories, and later at SRI International (formerly known as the Stanford Research Institute), on a classified program investigating psi phenomena for the U.S. government. On a much more personal note, he is also a warm and helpful man, whom I cannot thank enough for his permission to quote from his book. If you are going to truly study this science, please read it. My book is just an overview; The *Conscious Universe* is a true source of scholarly erudition. Much of this chapter is adapted from his book, and it is the source of all direct quotes in this chapter.

The term psi derives from the first letter in the Greek word *psyche,* which means *soul* or *mind.* Different varieties of psi exist, such as mind-to-mind communication (telepathy), predicting future events (precognition), affecting matter with your mind (psychokinesis), and the one that allows us to see distant objects and events—our favorite, clairvoyance.

If I were to describe psi research to a layman, I would say it is research into out-of-body consciousness. But does out-of-body consciousness exist? Certainly, you can't have an afterlife if you aren't conscious outside your body. So, can we absolutely prove psi exists? The answer seems to be yes.

From the late 1920s until 1965, Professor Joseph Banks Rhine worked at Duke University testing psychic abilities. Rhine developed the well-known deck of cards that uses five symbols (square, circle, wavy lines, star, and triangle) on five cards each. His methodology was to have one person act as the sender, who would select the top card from a shuffled deck, and try to send that symbol mentally to somebody some distance away, the receiver. The receiver would then announce which symbol he thought the sender was looking at.

With five different symbols, somebody who was just guessing has a one-in-five chance of getting it right each time, and thus the average result should be five right out of every twenty-five-card deck. Rhine's results always came up with a better average than that. You would also expect some of those results would be less than five out of five if it was pure chance, but that never happened. Something was positively influencing the results.

Rhine's tests have been tried thousands of times, and the results have consistently supported the existence of psi. So that's that, right? Psi exists; let's change the metaparadigm.

Unfortunately, regardless of all these individual studies, the results remain subject to a variety of scientific criticisms. There could be selective reporting, design flaws in the experiments, or even what is called *sensory leakage* (hints given subconsciously by the scientist to nudge the subject into the right answers). One advantage of all this skepticism, however, is that it "refined the methods used in future experiments"; the experiments continued to meet higher and higher standards.

One thing that became evident over time is that positive results could be attributed to clairvoyance rather than telepathy, which meant that a sender was not needed.

To address this, a new experiment conducted from 1966 to 1972 at Maimonides Medical Center in Brooklyn had senders attempt to transmit images to a sleeping receiver in the middle of a dream. Theoretically, the receiver would "incorporate those images into the dream." That's a fairly drastic redesign of the experiment, isn't it? But wait, there's more.

In the mid-1970s, researchers designed a telepathy experiment using a sensory deprivation technique called *ganzfeld* (German for *whole field*). The idea was to limit the amount of sensory input in order to eliminate noise, "thereby improving the likelihood of perceiving faint perceptions that are normally overwhelmed by ordinary sensory input." This experiment was performed in ten different laboratories and the results were significantly better than chance. Radin says "We now know that . . . we are fully justified in having a very high confidence that people sometimes get small amounts of specific information from a distance without the use of ordinary senses." If we have a high degree of confidence that we can receive information, could we also send information? Radin offers this:

If we consider all the ESP card tests conducted in laboratories all over the world, the likelihood that all of these experimenters would arrive at consistently better-than-chance results is phenomenally small—"a billion trillion to one." That means there's a billion trillion to one chance that ESP *does* exist, with "one" meaning "they just got lucky." So did they just get lucky? I think not.

Even the U. S. government has performed these experiments and gotten some amazing results. Here's a story Radin tells about the government experimenting with remote viewing, demonstrating how plausible clairvoyance actually is.

> Sometimes the results were so striking that they far exceeded the effects typically observed in formal

laboratory tests. In one test conducted at the request of government clients who wished to see how useful remote viewing might be in real intelligence missions, Dr. Edwin May described how a remote viewer was able to successfully describe a target, having no prior information about the target other than that it was a "technical device somewhere in the United States." The actual target was a high-energy microwave generator in the Southwest. Without knowing this, the "viewer" drew and described an object remarkably similar to a microwave generator, including its function, approximate size, and housing, and even correctly noted that it had "a beam divergence angle of 30 degrees."

Most of the classified, mission-oriented remote viewings could not be evaluated as controlled, formal experiments, because that was not their intent. In some cases, however, unexpected information obtained through remote viewing was later confirmed to be correct, and this was important because it demonstrated the pragmatic value of this technique for use in real-world missions.

So, how likely was it that the remote viewing experiments were just lucky guesses? They calculated that probability too, and the likelihood that the remote viewing experiments were legitimate, not just luck, and was "more than a billion billion to one."

Here's what Radin says about what he calls "anomalous cognition"—essentially, psi—and the likelihood of its existence: "It is clear to this author that anomalous cognition is possible and has been demonstrated. This conclusion is not based on belief, but rather on commonly accepted scientific criteria. The phenomenon has been replicated in a number of forms across laboratories and cultures. . . . I believe that it would be wasteful of valuable resources to continue to look for proof."

I especially like that last part—Radin's saying that we should just accept that psi exists and stop throwing our money away on further research intended to prove it.

Radin concurs that the explanation for psychic abilities is found in quantum theory. He says that such abilities "exist outside the usual boundaries of space time," which basically means that they're nonlocal.

He also examines intention, or as he puts it, "collective wishing." I include this because it is a real-world example of how scientists look at reality and because its results coincide with Jahn and Dunne's experiments with random-number generators.

> To test whether collective wishing made a difference, Nelson examined the historical weather data for the days before, during, and after graduation at Princeton University for a period of thirty years. He paid most attention to the daily precipitation data recorded in the Princeton, New Jersey, area, and in six surrounding towns, which acted as "control" locations. He predicted that on the day of graduation there would be more sunshine and less rain in Princeton than on the days before or after.

Nelson's analysis revealed that on average, over thirty years, there was indeed less rain around graduation days than a few days before and after graduation, with odds of nearly twenty to one against chance. An identical analysis for the average rainfall in six surrounding towns showed no such effect. Over thirty years, about 72 percent of the days around graduation had no rain at all in Princeton, whereas only 67 percent of the days in the surrounding towns were dry.

Curiously, on graduation day itself, the average rainfall was slightly higher in Princeton than in the surrounding towns, owing to a massive downpour of 2.6 inches on June 12, 1962. The average rain in the surrounding towns on that same stormy day was only 0.95 inches. What makes this even stranger is that the members of the Princeton Class of '62 reported that the massive rain that day held off until after the ceremony had ended! As Nelson then pointed out, this study prompts us to reconsider the old witticism, "Everyone talks about the weather, but nobody does anything about it."

Psi exists, and we can use it, but there's going to be a long laundry-list of details about it that we need to iron out to use it effectively. But as Radin points out, we can easily split the uses of psi into five categories: medicine, military, detective work, technology, and business.

We know clairvoyants are used in police work; just look at all the TV shows. We know they are used by the military; the Japanese are using them in business; and psychic warfare is talked about in one of the oldest Chinese military books.

Are you aware of Sun Tzu's *The Art of War*? Even *he* talks about uses of psi, or as it was known back then, chi (pronounced *key*). These soldiers were literally able to conduct psychic warfare on their enemies on a psychological level.

The National Security Council (NSC) has used remote viewing, but they were not the only agency. In September of 1979,

> The National Security Council asked one of the most consistently accurate army remote viewers, a chief warrant officer named Joe McMoneagle, to "see" inside a large building somewhere in northern Russia. . . .
>
> Still, because McMoneagle had gained a reputation for accuracy in previous tasks, they asked him to view the future to find out when this supposed submarine would be launched. McMoneagle scanned the future, month by month, "watching" the future construction via remote viewing, and sensed that about four months later the Russians would blast a channel from the building to the water and launch the sub.
>
> Sure enough, about four months later, in January 1980, spy-satellite photos showed that the largest submarine ever observed was traveling through an artificial channel from the building to the body of water. The pictures showed that it had twenty missile tubes and a large, flat deck. It was eventually named a *Typhoon* class submarine.
>
> Scores of generals, admirals, and political leaders who had been briefed on psi results like this came away with the knowledge that remote viewing was

real. . . . the U.S. Army had supported a secret team of remote viewers. . . . those viewers had participated in hundreds of remote-viewing missions, and that the DIA, CIA, Customs Service, Drug Enforcement Administration, FBI and Secret Service had all relied on the remote-viewing team for more than a decade, sometimes with startling results .

The military is still pursuing uses of psi—Radin points out how complicated modern fighter jets and the tactics necessary to use them in effective dog fighting are. It is estimated "that about five percent of fighter pilots have accounted for about forty percent of the successful engagements with hostile aircraft . . . in every aerial combat since World War I."

Radin comments on how "psychic" some of these ace pilots must already be, saying that, "there is some sixth sense that a man acquires when he has peered often enough into a hostile sky—hunches that come to him, sudden and compelling, enabling him to read signs that others don't even see. Such a man can extract more from a faint tangle of condensation trails, or a distant flitting dot, than he has any reason or right to do."

In this part of psi, what you believe comes true. In a sense we are creating our own realty.

Søren Kierkegaard said "There are two ways to be fooled. One is to believe what isn't true; the other is to refuse to believe what is true." Any new science—such as the *science of reincarnation*—has a serious hurdle to overcome when it is first introduced. Against any new anomaly discovered in scientific research, people will put up a barrier of skepticism because it threatens to change their world, and "to function without the annoying pain of cognitive dissonance, groups will use almost any means to achieve consensus. . . . This means that in the initial stages of a new discovery, when a scientific anomaly is first claimed, it literally cannot be seen by everyone. We have to change our expectations in order to see it."

This scientific hurdle—where a theory has to be believed before it is accepted, and then believed again—is part of a self-fulfilling prophecy. If you believe it, it will happen. Radin tells us of an interesting experiment in self-fulfilling prophecy:

> An experiment demonstrating the self-fulfilling prophecy was described by Harvard psychologist Robert Rosenthal in a classic book entitled *Pygmalion in the Classroom*. Teachers were led to believe that some students were high achievers and others were not. In reality, the students had been assigned at random to the two categories. The teachers' expectations about high achievers led them to treat the "high achievers" differently than the other students, and subsequent achievement tests confirmed that the self-fulfilling prophecy indeed led to higher scores for the randomly selected "high achievers."

Such studies make it absolutely clear that when experimenters know how participants *should* behave, it is impossible not to send out unconscious signals.

Of course, the most massive scientific hurdle here is the fact that any sort of parapsychological theory doesn't fit the preexisting scientific metaparadigm. Radin, too, makes a call to arms for changing this metaparadigm, saying that when the "evidence for an anomaly becomes overwhelming, and the anomaly cannot be easily accommodated by the existing scientific worldview, this is a very important sign that either our assumptions about reality are wrong or our assumptions about how we come to understand things are wrong. Or perhaps both are wrong. Assumptions at these fundamental levels act as extremely powerful drivers of expectation and belief, and as we've seen, we only see what we expect to see." We already have a lot of evidence pointing to some very intriguing phenomena that can't be explained with the current metaparadigm. Psi is staring us right in the face, but you have to believe it in order to see it.

Radin adds to the argument against our current metaparadigm by pointing out the case in which nonlocality was proven true, which I'm sure you all know by now flies in the face of common sense. But what we think of as *common sense* is a product of our current outdated metaparadigm; nonlocality does exist and there's no use denying it. It took only a "handful of experiments" to verify nonlocality. Radin shares my dismay—it only took a small amount of compelling data to prove nonlocality and get it put into textbooks, but these same minds are keeping psi far away from our scientific journals. Why?

It's because our current metaparadigm, quantum theory included, allows for anomalies such as nonlocality, but "hardly anything predicts psi," even though we have such massive evidence for it.

Let's not lose all hope, however. Radin points out that we may have a way of bringing psi into the scientific mainstream—we just have to put it through the same paces nonlocality went through: namely, quantum theory. Radin says that "some scientific developments in recent years suggest a way of thinking about psi that is also compatible with mainstream scientific models. Four such developments are related to quantum theory. All four run counter to common sense, all four were thought to be theoretically possible but practically untestable, and all four have now been empirically proved. Of principal importance here is that all four *must* also be true to be compatible with what we know about psi."

He then points out how we can retroactively interpret other experiments when we have psi and nonlocality at our disposal—suddenly, little quirks in certain experiments, certain unexplainable defects, become clear. It's like finally figuring out what that extra key on your key ring unlocks—it's that sort of epiphany. Radin says:

> Interpretation of existing theories may change when viewed in the light of psi and nonlocality. For example, in the late 1980s, neuroscientist Benjamin Libet conducted an experiment in which he asked his subjects to flex a finger at the instant of their decision. He monitored their brain waves to see if the instant that the decision was made would be reflected by a change in brain waves. On average, the volunteers took about a fifth of a second to flex their finger after they mentally decided to do so, an expected time lag for the brain to activate the neuromuscular system. But according to their brain waves, their brains also displayed neural activity about a third of a second *before* they were even aware that they had decided to move their finger!

Libet interpreted this result as evidence that our sense of free will in deciding what we do may be unconsciously determined *before* we are consciously aware of the decision. If mental intention, which is connected to our most intimate sense of personal expression, actually does begin in a part of the brain that is outside our conscious reach, then perhaps *all* our behavior is completely determined by processes outside our control.

All of our psi questions can be answered in the world of quantum mechanics; this is how we can get psi into mainstream science. In the meantime, we seem to have all the evidence we need to get psi widely accepted. Dean Radin says, "As some of the stranger aspects of quantum mechanics are clarified and tested, we're finding that our understanding of the physical world is becoming more compatible with psi."

And, as many other scientists say, Radin also wants you, as part of the general public, to demand psi-related research and more of it. If there is public interest, then there will be funding and the attention of top scientists. Otherwise, we may very well see progress on these life-changing experiments "measured in half-centuries or centuries." If you're on the fence about supporting psi research, then stop and think about it. While you're thinking about it, notice that you're not *doing* anything. You're just shuffling thoughts through your head. Radin challenges us to ask ourselves, *are we really only physical beings? Really?* He quotes Carl Jung, who said "it is almost an absurd prejudice to suppose that existence can only be physical. As a matter of fact, the only form of existence of which we have immediate knowledge is psychic [i.e., in the mind]. We might as well say, on the contrary, that physical existence is a mere inference, since we know of matter only in so far as we perceive psychic images mediated by the senses."

There must be a massive push if we are to get funding for this research, though. Do you really think that current science wants to overhaul itself to include psi as part of its curriculum? Do you really think that our religious leaders, especially the more fundamentalist ones, want to admit that humans may have psychic powers? The problem with psi research is that it's too radical. Even though we have mountains of evidence that says it's true, we have to fight as Galileo fought for the theory of heliocentricity. This fight is not going to be easy, but just think of the world once we're able to redefine it.

So can we be certain that this information is true? The scientists that I quoted here think so. I want to end this chapter with a quote from an e-mail Dean Radin sent me "The likelihood that the reported ESP test results were due to chance is essentially zero."

CHAPTER 9

QUANTUM BIOLOGY
SECTION 3 MICRO PHYSICS

If the soul is a disembodied consciousness that exists in a wave form, then before we go on we should look at the reality in which it exists. To do so, I have to explain the terms *zero point field*, *zero point energy*, *nonlocality*, and *our essence*. At some point in this chapter, many readers might become confused. Don't worry, so are many of the greatest scientific minds of our time. You'll be in good company.

So let's simplify this.

Every particle, even the smallest you can imagine, exists as a wave. That includes the particles that make up you. You are a sea of waves. This pulsing ocean of waves exists all around us and we are a part of it. We know it is there because we can detect it. But the instruments we are using today are too crude to see it fully.

Imagine, if you will, that you are Alexander Graham Bell, the inventor of the telephone. You have just sent a message to Watson, your assistant in the next room. "Watson, come in here," you say. That one sentence is historic; it was the first telephone conversation ever. Today, just over one hundred years later, we transmit waves of light that carry digitized sound waves of your voice down fiber optic cables and off satellites, and then we turn those waves of light back into sound waves so that your mother in the next town can hear the sound of your voice.

With the science I am telling you about, we are at the stage of development that Alexander Graham Bell was at one hundred years ago. What type of science am I talking about? Everything communicates in pulsing waves of energy, and only when we observe something does it take a particle-type of form. This includes you.

Matter at its most fundamental level cannot be divided into independently existing units. Subatomic particles aren't solid little objects like billiard balls, but vibrating and indeterminate packets of energy that cannot be precisely quantified or understood in themselves. Instead they behave sometimes like particles; a set thing confined to a small space, and sometimes like waves, vibrating and diffuse, spread out over a large region of space and time. Sometimes they behave like both a wave and a particle at the same time. You could call them wavicles if you like.

Because subatomic particles can act as waves or particles, or even both at the same time, they have an effective range far outside that of normal particles, in that one particle no longer has to be touching another particle or exerting any energy on it whatsoever in order to affect it. This is called *nonlocality*.

This describes the way a quantum entity such as an individual electron can influence another quantum particle instantaneously over any distance, despite there being no exchange of force or energy. It suggests that quantum particles once in contact retain a connection even when separated, so that the actions of one will always influence the other, no matter how far they get from one another. Albert Einstein disparaged this as "spooky action at a distance," but it has been decisively verified by a number of physicists since 1982.

In the last chapter, we saw how the human mind can reach through time and space, and theorized that we can extract information because of this wave/particle duality. This theory is based on the evidence from clairvoyant archeology and the research at Princeton University.

At this point in the book, I want to discuss the soul again, and provide a working definition for our purposes. The soul is really your disembodied mind, which retains its reference; that is, it's you. When you die, your mind, your awareness, continues, even though your body is no longer with it. For the sake of our discussion, if we find the mind, we find the soul. Your *consciousness*, *mind*, and *soul* are all synonymous. The terms describe the same entity in different locations. Just as H2O can be ice, fog, or steam, it's still all water.

If we are going to explain reincarnation, we need to explain how the mind interacts with the body, and we need to bring the discussion to a biological and quantum level.

I don't want to lose any readers with the term *quantum*. It simply refers to physics as opposed to chemistry; just as there is a chemistry to your body, so there is a physics to your body. How those two interact is how the soul gets in and out of your body. I intend to show, in easy steps, how science explains it. In particle form, you are chemistry; in quantum or wave form, you are a soul.

Keep in mind that you cannot be a particle without being a wave as well. You cannot have one without the other.

Although I'm going to talk about a couple of experiments and use some technical terms, the stories of the scientists themselves—what they went through trying to develop their experiments, the criticisms they endured, and their victories—are compelling human stories.

In 1987, Jacques Benveniste of INSERM, the French Institute of Science, conducted the following experiment. He took a common antibody, a substance that causes allergies, and exposed it to certain white cells in the blood called basophiles.

If you are allergic to a bee sting, molecules of bee venom would not be in your body more than a few seconds before they triggered basophiles to degranulate. This is one of the ways your body protects itself from invading poisons. In his experiment, Beneviste prepared a solution that was certain to trigger degranulation. He exposed the white blood cells to the antigen solution, and got the expected response: the white blood cells degranulated. Benveniste then diluted the antigen solution until it was no longer chemically active, but he still got the reaction. He continued to dilute the antigen solution until he had , essentially, nothing but water. Using a standard formula known as Avogadro's constant, he mathematically confirmed that it was impossible for the water to contain a single molecule of the antigen. When he exposed the basophiles to this extremely diluted antigen solution, they still degranulated.

That's not supposed to happen; his result was outside the parameters of known science. He repeated the experiment seventy times, and asked other research teams to repeat it in Israel, Canada, and Italy; they all came up with the same result.

His interpretation of the results was that the water that had contained the antigen retained an electromagnetic memory of the antigen that had been in it. But such a view is outside the bounds of orthodox science. Science could not explain this apparent instance of homoeopathy in Benveniste's research.

Benveniste was criticized for his findings, and others tried to demonstrate that his work was flawed, but the labs replicating his experiment were too tough a nut for the naysayers of the scientific world to crack. So other labs confirmed his result to the experiment, but disagreed with his interpretation of the result. This was really the beginning of digital biology.

If Benveniste's interpretation of his experiment was right, there must be other ways to prove it. And if that was possible, what exactly was being proven?

In the next experiment, a team of French scientists removed the heart from a guinea pig, and kept it beating (in doing this, they blurred the line between chemistry and physics).

While they initially used chemicals to keep the heart functioning, they were also able to use a transducer and computer to create low-frequency electromagnetic waves, keyed to the frequency of the chemicals, which could speed up the heart in the same way the chemicals would. In essence, the electromagnetic signature of the chemicals could do the same thing as the chemicals themselves. There was no difference between the chemical molecule and its frequency in terms of result. Consider this analogy. If you had a headache, you could either take an aspirin or listen to the electromagnetic signature of aspirin, and both of them would make your headache go away. Each would be equally effective. In essence, the French scientists discovered that molecules speak to each other in oscillating frequencies.

So, if everything is pulsating waves, from the chemicals within us to the world around us, we and our consciousnesses reside in an ocean of waves of pulsing energy that goes through us and interacts with us. Have we found scientific evidence for this? Yes, we have. It is called the Zero Point Field.

At any given point, where all matter has been removed, and there is a complete and perfect vacuum, and the temperature is absolute zero, we can still detect energy. In her book, *The Field*, Lynne McTaggert calls this zero point energy, but it could also be dark matter, because matter and energy are basically one and the same.

This energy pops in and out of existence, or at least in and out of our one temporal and three spatial planes. If they have a consistent existence, it is in a dimensional place we cannot access with our senses. This is just the physics of it, folks. But there is more.

These particles exist in wave form. They do not become particles until they are observed. If we apply this to human consciousness, it means we existed in wave form before we were born. In Zen Buddhism, there is a question asked of a monk. "Where do you go when you die?" He answers, "Back to the place I was before I was born." In a few chapters I will show how religious beliefs held across the world support the scientific findings we've been discussing, and how these scientific findings support doctrines of the world's religions. But for the moment, let's stay on the science.

Quantum physicists discovered that any quantum particle would collapse into a set entity when it was observed or when a measurement was taken. A quantum particle would react to observation and become a definite object. Remember how water never seems to boil when you're watching? It's as if the water didn't exist until you looked into the pot, at which point it would come into being.

Looking at these particles, observing them and measuring them, force them into a set state. Until we looked at them, these particles could only be considered as probably being at a point in time and space until our observation froze them in a set state.

Now think about how particles act and the nature of reality. It means that you, the observer, bring an object into being by observing it. Nothing exists until we observe it.

So if you were to put this book down and look around the room, everything exists around you as a wave form. That is its basic component. You are yourself a wave form, and when you look at anything in the room, it comes into being, because you are observing it.

If your wave form signature is limiting its perception to perceive other wave form signatures as materiality, and you view your spirit as being within you, as so many religions do, then what you've just read is an application of the standard scientific explanation for how we see to describe in scientific terms your soul and its perception.

In short, everything around us is in a wave form until we look at it, and then it appears to collapse into a particle/object.

If we can digitize chemicals and they perform the same as either chemicals or digitized oscillating frequencies, then can I digitize you? I know that some readers will immediately think of the transporter on Star Trek. Focus, people! If your mind can reach out through time and space in a form that is wavelike, then can that same mind retain its frame of reference without your body?

Doesn't it do that already when it reaches out through space and time, as we proved in the clairvoyant studies or in the Intention Experiments? Is that proof that your mind exists outside your body without losing its essence?

We have already seen that it can do this. So can it do this after your body dies? Does it need your body at all? What role does your body play? What role does your brain play?

In *The Field*, McTaggert says "Our brain primarily talks to itself and to the rest of the body not with words or images, or even bits or chemical impulses, but in the language of wave interference: the language of phase, amplitude and frequency... To know the world is literally to be on its wavelength."

She goes even further when she explains that we see with waves as well. Because we don't see objects in our brains, but in the world, she says, we must be projecting the images we see back out into the world, to perfectly coincide with the original object we're observing. "...we are transforming the timeless, spaceless world of interference patterns into the concrete and discrete world of space and time... the lens of the eye picks up certain interference patterns and then converts them into three-dimensional images."

So, your brain thinks with waves. It sees with waves, smells with waves, and hears with waves. Perception occurs not in the physical macro world, but on the level of quantum particles. We don't actually see physical objects; we see the quantum information that those objects send to us as waves. "Perceiving the world was a matter of tuning in to the Zero Point Field."

Scientists are trying to figure out how this occurs physically and mechanically.

If we're interacting with these interference patterns or energy patterns, which parts of our bodies pick up this information and turn it into the objective, concrete world that we see? McTaggert suggests the possibility that "the microtubules within the cells of dendrites and neurons might be 'light pipes,' acting as 'waveguides' for photons, sending these waves from cell to cell throughout the brain without any loss of energy."

According to this theory, microtubules and the membranes of dendrites represent the Internet of the body. Every neuron of the brain can log on at the same time and speak to every other neuron simultaneously via quantum processes.

Let's develop this theory. We don't just see images in the back of our brain; we perceive them in three dimensions and can even manipulate the images we project. In real time, we are creating our own world with just our observations. It's possible, then, that consciousness is "a global phenomenon that [occurs] everywhere in the body, and not simply in our brains. Consciousness, at its most basic, [is] coherent light."

This explosive discovery about quantum memory leads to the most outrageous idea of all: short-and long-term memory doesn't reside in our brain at all, but instead are stored in the Zero Point Field. (I will explain quantum memory in the next chapter.)

So, what is your brain, then? It should no longer be considered a storage bank—all your memories reside in the Zero

Point Field. Therefore, as some scientists, including Ervin Laszlo, would theorize, the brain isn't like a computer's hard drive anymore—it's more like a modem, receiving and sending information back and forth from external storage. That's an enormous change. And if our memory is in the Zero Point Field, a metaphysical Internet of sorts, what does that say about us as beings?

The data from the Intention Experiments demonstrate that the human mind, by will alone, can influence a program on a computer. It can also influence people. Coupled with the discovery by quantum physicists that subatomic particles react to almost any particle around them, that they are simultaneously particles and waves of energy, and that observation alone can influence them, the universe becomes a highly interactive, intimately networked system far beyond what common interpretation suggests. It allows the mind to reach beyond time and space to retrieve information. This separates the mind from your body, your momentary reality.

The case for reincarnation benefits from this redefinition of the mind's reach and ability. If the mind is able to reach out and influence other systems, then it is possible that a human's intellect—the soul—can travel beyond the physical brain and live on after the body is deceased.

So, if the location of the mind is so elusive, could we locate where memory is stored, as well as how it's stored? The theory will surprise you.

CHAPTER 10

THE MYTH OF THE HUMAN BODY

There are two points this chapter makes while giving a little history of how we got here.

Point 1: If your true essence is a conscious wave form whose consciousness transcends death, then why don't we know it? We have proven that psi exists, convincingly and scientifically. No one can tell you why or how you have gotten a conscious life but we can tell you that everything exists in waves. You interpret other people from your perception of their wave form because everything is packets of energy we perceive as materiality.

You "see" reality in a constricted manner. Your vision only picks up certain wavelengths of light. You hear in a certain frequency range.

Point 2: Your consciousness is made up of a wave form that is more permanent than your body. The matter in your body is impermanent not only in death, but in life as well. We are constantly replacing the matter in our body, swapping one bit for another.

So let's start.

Karl Lashley is a neuropsychologist who spent many years investigating where in the body memory is stored. His methodology was to train rats to do various tasks, like negotiating a maze. After they'd demonstrated that the task had been learned, he would surgically remove different parts of their brains. (Crude, and possibly cruel, but this was the forties, before scientists worried about things like that.) He assumed that he would eventually find exactly the portion of the brain where the learned tasks were stored.

Instead, what he found was that no matter what part of the brain was removed, the memory stayed intact. Even when he took out large parts of a rat's brain, as long as they had motor skills left, they could still drag themselves through the maze, the way they had learned to.

This made no sense if memory had a specific storage location. It seemed that instead of being stored somewhere in the brain, memory was stored *everywhere* in the brain. When a colleague of Lashley's, a younger neurosurgeon named Karl Pribam, read an article in the mid-sixties on the first holograms, he had a flash of insight, and for the first time thought he could put together a theory to explain Lashley's rats.

Holography is a product of *interference*, the crisscrossing pattern that's generated when two waves intersect. For example, if you drop a pebble into a pond, it creates a series of concentric waves that spread outward. If you drop a second pebble in, the second set of waves will expand and the two sets of waves will pass through each other. The pattern this produces is an interference pattern, anything that takes the form of waves, like light waves or radio waves, can generate an interference pattern. Laser light, an extremely coherent form of light, is particularly good at creating interference patterns. So here is how you create a hologram. You take a laser and run the light through a beam splitter, creating two separate beams from the same beam of light. Using mirrors, you bounce one of the beams off the object you wish to make a hologram of, and then direct that beam through a holographic plate.

The second beam of light is bounced off another mirror toward the same holographic plate. The two beams create an interference pattern on the holographic plate, like the pattern those two pebbles make in the pond.

The image on the film looks nothing at all like the original object. It looks like sets of concentric rings. But when another laser beam is directed through the film, a three-dimensional image of the original object appears behind it.

You can walk around a holographic projection and observe it from any angle, just like a physical three-dimensional object. But the thing that is even more amazing, and what got Pribram all worked up is this. If you cut the holographic film in half, and shine a laser through it, each half will project the entire image. If the halves are divided again, an entire image can still be created from each portion of the film. The image will get less and less distinct, grow fuzzier with each division, but the entire image is contained in each piece of the film.

Let me say that again. Unlike normal photographs, every small fragment of a piece of holographic film contains all the information recorded in the whole.

This is why Pribram got so excited, because he finally saw a model that explained how memory was stored. If a fragment of holographic film could contain all the information contained in the original image, why couldn't every part of the brain contain everything to recall a whole memory? It would explain why the rats could remember the maze when they had parts of their brains removed. Portions of their brains retained their whole memories just as portions of the hologram retained the entire image of what had been photographed.

So, do we have evidence of this kind of thing at the macro level?

As reported in *The Guardian*, a woman named Claire Sylvia underwent a heart and double lung transplant in 1988. Afterward, she began to crave foods she had never cared for, found herself attracted to women, and started having dreams about a man named Tim. She hunted down the family of her donor, who turned out to be a young man named

Tim, and her cravings were for all his favorite foods. This type of effect is called transplant memories, the heart and lungs carrying memories of its previous owner.

What does this show? Your memory is stored holographically throughout your entire body. And since your mind must be able to access your memory, your mind must be able to operate throughout your entire body.

But this still needs a further explanation. If mind and memory are diffused wave patterns, then there has to be a theory of how they interact with our bodies on the macro level.

Apparently, we are not there yet. There is no well-developed, sophisticated theory to explain how this fits in our current understanding of biology. We measure the electromagnetic pulses of the heart in an electrocardiogram, and we're always improving our imaging of brain scans, but the energy patterns of the body itself are only at the early stages of being mapped, and it is these energy pulses that create the same kind of interference patterns that you see in a hologram. These energy pulses also create auras, which we're able to capture on certain types of film, and it is these auras that can reach out and intersect other energy pulses that everything around us emits. We are, in essence, reducing our bodies into simple energy for such analyses.

In the 1600s, blood was nothing more than blood, and when a man was ill, it was because he had some bad blood. The doctors cut him and drained out the bad blood. Today, we know that blood is not merely some red liquid coursing through our bodies that needs to be drained when we have any sort of disease. It's a life-giving solution rife with white cells, red cells, plasma, leucocytes, and all sorts of things that are carried through our body. Today, our understanding of our body's energy patterns is just as primitive as the seventeenth century's ideas about blood. Three hundred years from now we may have a more detailed knowledge of the body's energy patterns, because right now we are finding that these energy patterns can intersect and interact with other energy patterns at the quantum level, resulting in the ability to carry vast amounts of information.

Scientifically, then, how do we separate mind and body for the point of this discussion? If our memories are stored holographically throughout our bodies, how do we separate mind and body?

At the beginning of this book I said that I would present the science of reincarnation to you, and let you decide for yourself if I have made my case. In study after study, I have shown that the human mind has far more ability and power than previously thought. At this point, you must now choose between two scientific models.

The first model is the classic model of the last few centuries. In this model, the mind is the result of our chemical bodies—in scientific terminology, the mind is a result of bio-electrical postulates. If you are person of faith, your belief in a soul will be based on your particular religion. There can be faith in a soul, but there is no empirical evidence of a soul.

The second model is the one I am advocating—that your soul or consciousness came to inhabit your body by means of the scientific processes I have described. After passing from your body, your soul/consciousness will still be you, with your memories and thoughts from this life intact, and will continue in a discrete form.

So I ask you again: which model is more consistent with the science?

I'm sure you're already familiar with the fact that your body regenerates. Let's reflect for a moment on just how dramatic this process is. Actually, 98 percent of the atoms in your body were not there a year ago.

Your stomach lining is replaced every day. You replace your skin every two weeks; the molecules and atoms in your bones are replaced every year. Even the enamel in your teeth is totally replaced every two years. Your body that existed three years ago is gone. And yet, you remember the taste of chocolate ice cream. Your memories from forty years ago are still there. So which is more permanent, your consciousness or your body? We are not talking about speculation here; body replacement is accepted as scientific fact. The body you had two years ago is no longer there. Every atom has been totally replaced.

What does this have to do with reincarnation? A critical argument is built in small, sure steps. According to Deepak Chopra, one of the best-known medical doctors in the U.S., a respected endocrinologist, and an expert in alternative medicine, my body does not create my mind, but my mind creates my body. In a sense, we reincarnate every year. The only part of us that stays the same are our thoughts, memories, and mentality—these constitute the permanent system that we build upon. As Dr. Chopra says, we are not "physical machines that have somehow learned to think," but instead are "thoughts that have learned to create a physical machine."

Chopra states his case with an analogy involving a magnet. Have you ever seen a diagram of magnetic fields and noticed how those fields curve around the magnet, then out and away at both poles? Chopra points out that these fields are invisible until you do the old grade school trick: rest a piece of paper on the magnet, and sprinkle iron filings on the paper. The filings will trace the curve of the fields, even though the actual fields are still invisible.

So are the fields that compose your mind-body activity. Chopra says, continuing that analogy, that "The iron filings moving around are mind-body activity, automatically aligning with the magnetic field, which is intelligence....And the piece of paper? It is the quantum mechanical body, a thin screen that shows exactly what patterns of intelligence are being manifested at the moment."

Chopra points out that the paper is a necessary part of the experiment. If you ever place a magnet near filings without the paper, you've got a mess—they cling to the magnet and you can never seem to clean all of them off.

Chopra says that the paper represents a gap between the body-mind (the filings) and the magnet itself, the *hidden intelligence*. We see the effect—the iron filings following the magnetic field that's under the paper, moving wherever the magnet does—but we don't see the cause, namely, the magnet. Our hidden intelligence remotely controls us, but because of the small gap between it and our body-mind, we can't see it.

There is an adage that you have to believe something before you see it. Chopra suggests that our brains are shaped in this way, and that we may be under-developing and handicapping some innate human parapsychological skills because we do not nurture them enough, even though nature intended them to exist and develop.

Chopra uses a case with kittens and their optic nerves to demonstrate this.

Kittens can't see when they are born. Their eyes are shut and their optic nerves have not yet developed. Their eyes actually learn to see in the first two or three days of their lives. Laboratory experimentation has shown that blindfolding kittens during these few crucial days will render them blind for life. Despite having all the physical equipment necessary to see, and a world full of things to perceive, the kitten will never perceive those things if during the early learning period for sight, the neurons of their brain and their optic nerves learned instead not to see. Sound familiar?

Chopra says that when the mind shifts, the body cannot help but follow. You can reshape your body by reshaping your mind. In another experiment he talks about, he shows how perception dominates reality, and that whatever *reality* may be, it is our perception that defines our world. Two psychologists, Joseph Hubel and David Weisel, took three groups of kittens, and during that critical period when their eyes were learning to perceive the world around them, put them in three distinctly different visual environments. One group was exposed to an environment consisting of horizontal black and white stripes, one group was exposed to vertical black and white stripes, and the final group was surrounded by plain white surfaces, with not features at all.

For the rest of their lives, the first group had trouble making out anything that wasn't horizontally oriented—for example, they were always running into table legs. The second group, the one that was exposed to vertical stripes in their first few days, had the opposite problem; they had trouble seeing anything that took the form of a horizontal stripe. The group that was raised in a featureless white environment experienced total visual disorientation; they had trouble making out objects of any kind. "These animals became what they saw," Chopra writes in his book *Quantum Healing,* "because the neurons responsible for sight were programmed into a rigid frame. In the case of humans, too, the brain sacrifices some of its unbounded awareness every time it perceives the world through boundaries."

Our consciousness is not, biomedically speaking, a part of our body. Any doctor can tell us how dendrites firing in our brain will send an electrical pulse down our nervous system that will cause, for example, our fingers to move. What these doctors cannot tell us is how the thought—"I want my fingers to move"—actually causes the dendrites to fire.

So, if you tell yourself that your mind is the result of your body, then for you that is true. You are creating your own reality. But you also have the ability to create another reality. We have already shown that you can change the world around you simply by intent. Simply by intent, you can bypass the limitations of your body's perceptions and create a new reality for yourself.

You still may say that all this can't be true, so for you, that statement is true. But for me, it is not, because I am creating a different reality for myself. I have already proven to myself that it is possible. It has been proven for me by the scientists I have already discussed.

So, if my intent is to create an afterlife for myself, then I would design a place that's really cool. Am I, then, a co-creator with whatever deity I choose to worship? The science says that I get the same effect no matter which deity I choose.

If I knew with certainty my death would not mark the end of my life but rather was a transition to a different realm, then how I experienced this life would change. Life would not have the same sense of urgency; it would not affect me in the same way.

If water retains electromagnetic memory and if our mind is a wave form, then our bodies are something our mind peers into and through in order to experience a particulate world or reality. Our bodies are transitory, as we have proven, but what are permanent are the mind and the memories temporarily housed there.

It is like life on other planets—the universe teems with life. Life is not a single event that only happened here on earth.

The variety of life on earth is endless. We are finding microscopic life on the other planets (e.g., Mars) and on the comets that fall to earth. No single event occurs; they all repeat, as we will see in the next chapter on fractal geometry.

So if memory is stored holographically, where is my consciousness? It is either in the wave form or in the chemicals we are made of. The limitations of our own bodies preventing us from seeing the world in wave patterns is what allows us to perceive this reality.

Before I leave this chapter I want to ask you a question. I gave you two models to choose from. Which one do you think is right? Do you think a bunch of chemicals that change totally every few years produces your consciousness? Or do you think your consciousness is more permanent than your body? Enquiring minds want to know.

I can't leave this chapter without recommending two books that were very helpful in writing it, *The Holographic Universe* by Michael Talbot and *Quantum Healing* by Deepak Chopra. On my web site www.thescienceofreincarnation.com you will find a link to an interview with Mike Talbot and Jeffery Mishlove.

CHAPTER 11

FRACTAL GEOMETRY

I promised at the beginning of this book that I would make it easy to understand, and I am going to keep my word. To do so in this chapter, I must go straight to the conclusion. No science can be complete without mathematics, and the fact of the matter is that mathematics supports reincarnation. The math I am talking about is *fractal geometry*. So now that you know the conclusion, you don't have to read this chapter. But if you do read the rest of the chapter, I am still going to make it interesting, entertaining, and easy to understand.

Fractal geometry has two main points, iteration and self-similarity. I am going to explain each by using an example.

Iteration: consider one drop of water from a leaky faucet, then another, then another, then another, then another, then another.... Each drop is one iteration, but in fractal geometry, there is never one, there are many. Fractal geometry is about the multitude.

Self-similarity: each drop of water is similar to the others—not identical, but similar. The drops can grow bigger or smaller or do both over time. They can be filled with rust (iron oxide), which makes them brown. They can be milky or not. But they are all self-similar. That's everything you need to know about fractal geometry.

So let's see how you interact with fractal geometry.

Fractal geometry is the visual side of mathematics. Did you see the three most recent *Star Wars* movies, in which the computer graphics were much more realistic than the special effects—the models, puppets, post-production laser effects—of the original trilogy, particularly when it came to designing planets? Fractal geometry allows computers to model mountains and lakes realistically to achieve stunning visual effects. Fractal geometry was practically applied in these movies.

Classical mathematics could describe the forms that humans could make—rectangular buildings, water towers, etc. In the 1970s, a mathematician named Benoit B. Mandelbrot showed how fractal geometry could describe the forms that nature creates—leaves, trees, plants, clouds, and weather systems.

Fractal geometry involves taking an equilateral triangle and then, on each side, creating another equilateral triangle;

then on all of the sides of the new triangles creating more equilateral triangles, etc. Each iteration becomes smaller and smaller while extending the length of the perimeter.

The object created has self-similarity—triangles upon triangles upon triangles.

How does this fractal geometry manifest itself in nature? If we measure one tree in a forest, we can find that the pattern of branching is similar from the base of the tree all the way out to the tips. If we measure the circumference of the trunk and take it all the way out to the circumference of the tips and then take a census of the forest, the forest will have several self-similar trees—the sizes of the trees in the forest match the sizes of the branches in the trees. In short, we see the iteration of fractal geometry going from the tree to the branches and from one tree to several trees. By using fractal geometry we can by measuring one trees height circumference and branching pattern deduce the size and distribution of all trees in the forest. This self-similarity is seen throughout nature. What was once a tangled forest now has mathematical symmetry.

Physiological processes are also fractal. For many years we thought that the beating of a heart was very regular. Galileo measured the swing of a pendulum by holding his finger to his pulse. Then, as we developed more precise instruments, we found that heart rates fluctuate, and that a healthy heart rate has a fractal architecture.

Similarly, eye trajectories, when plotted, have a fractal architecture. The eye does not look at something and absorb it, but instead moves randomly about it, from it, and back to it—the eye's movement is fractal.

Fractal architecture in antennas allows progressively smaller antennas to pick up a progressively wider range of frequencies. Every cell phone has an antenna with a fractal design. Without this fractal design, we would need a separate antenna to pick up all the different frequencies of waves that WiFi and Bluetooth would need. A fractal antenna can pick them all up.

Now what does this have to do with reincarnation? First, there is the self-similarity of the iteration. You cannot deny your own consciousness, but you are only one iteration and nature does not produce just one of anything.

Just to be clear about this. The self-similarity of an iteration does not mean something is identical. No two snowflakes are alike. They are similar. Nature produces one class of things, like humanity. There are no two truly identical individuals, but within this class, primates, there are other iterations of primates.

The second is self-similarity, which means consciousness mirrors itself. In short, the soul, like you, is the sum experientially of the total lifetimes it has lived.

Fractal Geometry

Inferring the Whole

What do we mean by inferring the whole from seeing a few parts of the puzzle? A child comes home from school and sees a pie pan on the kitchen counter. There is one piece of apple pie in it and some crumbs around the pan. He can infer from seeing the one piece that there was once a whole pie in that pan. He can also infer that his father had a large piece and his mother had a smaller one and his siblings had pieces with ice cream.

With this mathematics, all the evidence—the anecdotal information we have on past-life memories from the children in the University of Virginia studies, the near-death experiences people have had, and the past-life regressions they report—is no longer unverified but becomes both statistically predictive and a form of proof.

By living life we have proof that people live one life. We cannot infer from that more than one event. Living one life is one iteration for a soul. If a soul is to live many lives, then one life is only one event. In the case of the children studied at the University of Virginia, we see two iterations. But with past-life regression, we can see ten or more iterations. With near-death experiences, we see typically two, and sometimes three, iterations. How could this mathematics possibly take these events from being anecdotal in nature to a form of proof?

In these types of events, we see a global consistency. It is a form of the gold standard of scientific research—replicability. These descriptive events, even if we do not understand the cause, are consistent across all human groups, regardless of sex, age, race, creed, or location. They meet two important criteria in making reincarnation a true science. First, they support the theory we have already established regarding how reincarnation may operate scientifically, that is, on a quantum-mechanical scale. Secondly, they give us a method by which to predict the future, something a scientific paradigm requires.

But does the human experience of reincarnation stand alone in the animal kingdom? This question may seem impossible to answer, but in fact we have evidence to the contrary in the migratory patterns of the monarch butterfly.

Monarch butterflies start their migration from Mexico. The monarch butterfly sanctuary is in Angangueo, Michoacan Mexico. The first leg of their migration is to the middle of the United States. There, that generation of butterflies dies. The next generation flies to Canada, where that generation dies. The monarch butterflies that journey back to Mexico are the great-great-grandchildren of the previous generation that made the flight from Mexico.

Monarch butterflies that are migrating to Mexico from Canada have been captured in Kansas, flown to New York, and released. Upon their release, they fly on a heading as though they were flying to Mexico from Kansas. In a few hundred miles, they reorient their flight pattern and continue on to Mexico.

No one knows what mechanism causes the monarch butterflies to retain information about a location that they have never been to, nor their parents, nor their grandparents. Clearly this example is not proof of reincarnation, but just as clearly information is being retained generationally.

Critics could argue that this is not proof of reincarnation. However, under the current metaparadigm this type of memory retention should not exist. What other memories are being retained by the butterfly? Clearly the butterfly retains this memory from one life to the next. The children in the University of Virginia studies are also retaining memories from one life to the next. So, do animals reincarnate too?

No one can answer that, yet the anecdotal record shows that people who have had near-death experiences or children who have claimed prior incarnations speak of seeing pets and other animals in the afterlife. In short, continued consciousness after death is a common feature of all life, not just humans.

This result would clearly fit with theory. All animals are subject to the laws of physics, they are seen continually in the anecdotal records, and we have the observational account of the monarch butterfly that also fits. All this information fits within the mathematical model of the fractal geometry of reincarnation.

Finally, this model eliminates anomalies. The current metaparadigm cannot fit the anecdotal and observational events that we are seeing into its model of our understanding. This new model's paradigm not only includes the anomalies that exist under the current models, it provides a method to predict future events.

Because of the repetitive nature of the description and the many ways it comes to us, we can infer not only the whole, we can also now bring into focus the shape of the time scope.

Self-Iteration

While the geometry of all of life is self-iteration—from the making of coastlines to the making of capillary systems, from blood-flow systems in your body to the information-flow dendritic networks in your mind—if you look at the scope of time (billions of years in the case of this universe) as a river of information, and view yourself as one iteration, can you not see that there was an iteration of you before and an iteration of you after? This logic speaks to you from so many different directions, such as the children at the University of Virginia and their not-yet-proven past-life regressions. But past-life regression fits the story line of the children studied at the University of Virginia, as well as the mathematical model of continued iteration/lifetimes.

At this point in our cultural history, you no longer have to take on faith that you will reincarnate, because as we develop and evolve as a species, we're seeing that the logic and the information for reincarnation has been around us for a long time.

Intentionality

Let's spend a moment on cultural beliefs. If we can use the process of intentionality to create events, like sitting in front of a computer to make it do what we want it to do, or to be able to do remote viewing, then humanity as a whole has a cultural belief in life after death. That intentionality creates the reality, and that reality can eventually be measured, even if we don't have the devices yet to do it.

This is where the rubber hits the road, the *uh-oh* moment. The mathematics of reincarnation begins to emerge. In short, the mathematics predicts reincarnation. All of these anecdotal stories brought together mean nothing unless they can be melded into a scientific theory. In short, it has to be reduced to mathematics—fractal geometry—in order to see where that will take us in fifty or one hundred years.

We've already talked about iteration and self-similarity. Consciousness in one time period is a manifestation of consciousness in other time periods, and the retention of memory (as exhibited by the children who remember prior life experiences) is a manifestation of those other life experiences when we extrapolate it.

Let's follow the steps for a moment. If a child has a memory of a past life and a memory of a current one, you are seeing self-similarity, because that child has a memory of two separate and distinct lives that retained coherence through death and rebirth.

I will tell you more about Alfred Wegener in Chapter 14. For now all you need to know is that he developed the theory of continental drift. Just as Wegener did not understand the cause of continental drift, even though he developed the theory, we do not currently understand the scientific cause of people who recall past lives during regression. But we have something he didn't—a mathematical model that *predicts* our outcome because we see it in all of nature. Where Wegener now becomes important is that in fifty years we may have discovered the cause for reincarnation, but for now you can know that the scientific model we are describing is consistent with the mathematical structure of nature as well as what we are seeing in our macro observations.

Statistics are democratic. If someone doesn't like that, then it is his or her problem, not mine.

So what is true in science? The results of controlled laboratory experiments are truth, and the gold standard is the double-blind experiment. We can believe the results that a variety of independent laboratories get when those results are consistent. We may not understand why we are getting certain results, and such will be open to some speculation and spirited scientific argument. It is undeniable, however, that if a variety of laboratories conduct the same experiment and get the same results using the same protocols, and we can reasonably rely on the information from those results.

This book addresses issues raised by the results we are seeing from a variety of laboratories around the world that contradict current, accepted theory, and it is here that you have the disconnect.

Accepting the new results and incorporating them to modify our current theories begins to have cultural ramifications. At this point the information in this book becomes explosive, both culturally and socially.

I challenge the imams, the priests, the rabbis, the wiccans, the ministers, the bishops, the pope—all of them—with this proposition: if you see it here on earth, you will see it in the afterlife, because science is indicating to us that consciousness is discrete. This means that you'll retain your identity after death in some form, and it is possible that you will be reincarnated, because that is what the studies at the University of Virginia show.

So, if there is a Jew, a Muslim, a Catholic, an Irishman, an Englishman, and a Hindu sitting at a table, there will be a Jew, a Muslim, and so forth sitting at a table in the afterlife.

If that is the accurate representation of what our culture has told us, regardless of which culture it is, and if that's what the science is showing us, then it renders moot the holy wars occurring right now in the Middle East, Northern Ireland, Malaysia, Cambodia, and any other place you choose to go. The various religions all believe they all are the one true religion, and they're right, to a point. But they're wrong, of course, when they say that the *other guys* are wrong. The new mindset that you will experience with this awakening—that all religions are correct—both liberates you and completely changes the human paradigm.

I would ask you, as part of this challenge, to prove me wrong, scientifically and factually, with results and not beliefs. Give me objective proof, not subjective conjectures.

Why do I mention religion now in the discussion of scientific theory? It is because all of our cultural beliefs are supported by the mathematics of fractal geometry. You cannot be promised life everlasting in a religious sense and ignore the empirical data supporting it. In short, this scientific theory supports any religion that claims either an afterlife or reincarnation. This would include all the denominations of

Christianity, Judaism, Islam, and Buddhism. As a matter of fact, a lot of Buddhist doctrine resembles subatomic particle theory. Hinduism, Shinto, Taoism—this scientific theory supports all of them with voluminous proof in observable experiences. Remember, we have seen scientifically that prayer works.

I wrote earlier about the power of prayer, and now I have told you about the mathematics of prayer, because this mathematics is the mathematics of intent. Continually wishing for something constitutes an iteration, and each iteration is self-similar.

Science could hurry this process of discovery if it began charting the iterations and time periods of incarnations for some of the subjects studied at the University of Virginia. There must be parameters that reincarnation cycles follow. Virtually everyone in this study has had two incarnations, but some cases may actually produce more. These candidates should undergo past-life regressions.

Dr. Weiss's subjects claim multiple incarnations. Those should be charted as well. Anybody familiar with fractal geometry knows that very interesting diagrams can be created with it. Such diagrams would allow us to develop typical incarnation patterns. I could say that once you begin to chart such an event, it takes on a life of its own, if you'll pardon the pun.

So, if we have observed data supported by scientific theory, as well as a newly discovered branch of mathematics, that support both the observations and the theory, it seems we have . . . fact.

Perhaps.

So who believes this? Can we rely on this? Is there another metric we can use? Who is using this and how are they using this information?

Even if this were the only case I could make for reincarnation, it would stand on its own. But what you have read so far is really only one aspect of the case; there is so much more evidence. We need to flesh it out with the hard, factual evidence that has been accumulated and examined. To understand its strength fully, we need to look at how we got here and at those people who are sensitive to the transmission of wave form information.

All of us have this ability. Every person who has heard the phone ring and knows it is a relative calling has tapped into that transmission of wave form information. But like musicians, some of us are naturally skilled.

In the next section of this book we will look back at the beginning of this type of research, meet the *stars* of natural psychic ability, and hear what they have to say.

CHAPTER 12

APPLICATIONS BUSINESS AND MILITARY

If psi exists and if we have all the other information that I have given you, are scientists actively exploring this science now? What does the landscape look like? Stanford University is one of the leading universities in the United States. Here's what went on there.

In 1972, Russell Targ cofounded the Stanford Research Institute program that investigated remote viewing and other psychic phenomena. While he was at Stanford, he also worked with the CIA, DIA, NASA, the Navy, and Air Force and Army Intelligence to investigate remote viewing as a spying technique. Not only was one of our most prestigious universities taking psychic research seriously, but the research was fully funded and used by our own government.

I am not criticizing our government for doing this. My point is that they believed in this effort and they had success when they supported it.

Again, the question is: what does this have to do with reincarnation? And again, the answer is that human beings do not fully understand the total reality they live in. We have many abilities that we do not know about, and clairvoyance is one of them, which the Stanford researchers used for their remote viewing program. If we can see across time and space with our minds, then perhaps our minds—our discrete consciousnesses—can also travel free of the body, thus facilitating reincarnation. Can you see how much of reality is hidden from us?

We interpret that reality only through our senses—sight, hearing, smell, sound, touch. It was not until the latter part of the twentieth century that we understood that elephants can communicate with each other at sound levels lower than our ability to hear, and it wasn't until we put sensors in elephant herds in Africa that we were able to begin to actually track that communication. Currently, there is an ongoing study to build an elephant dictionary whereby we can understand what they're saying to each other.

The question you should be asking yourself is: what else are we not seeing? Russel Targ's work has some answers.

First of all, after a career in which he was one of the foremost researchers into ESP, Targ has no doubts about our psychic abilities. He states unambiguously that he witnessed it occurring in his laboratory daily.

In a recent article he wrote for *Quest* magazine, Targ describes some of the incidents and findings that convinced him. During the 23-year SRI program, Targ documented remote viewers who read code words off files in a secure NSA site in Virginia; described a Chinese test of a nuclear weapon three days before it happened; and located a downed Soviet bomber in Africa, all of this from his California lab, and all confirmed by one government agency or another. This confirms the mind's ability to reach beyond its own body to extract information regardless of time or distance.

You could not have a better pedigree than Russell Targ for this work, and his successful research has been paid for and confirmed by the US government.

A critical part of Targ's study, and a critical part of most of the studies described in his book, *Limitless Mind*, is nonlocality. As I have shown, nonlocality is fundamental to reincarnation, but it is also fundamental to quantum physics. "The most exciting research in physics today is the investigation of what physicist David Bohm calls 'quantum interconnectedness,' or nonlocal correlations," Targ writes. Physicists have conclusively demonstrated that two tiny particles of light going in completely opposite directions will continue to keep their connection to each other. Even though they're traveling at the speed of light away from each other, it's as if they're sitting right next to each other. To determine how this is happening, we need a construct of what we cannot see; namely, other dimensions. At this point, I have a question for all scientists:

does believing in and conceptualizing a dimension create it? Consider, for example, complex Minkowski space.

Complex Minkowski space is a nonlocal mathematical model that takes into account the three dimensions of space we're familiar with, as well as an imaginary dimension that plots time. When Einstein wanted to describe relativity, ordinary geometry couldn't do it, it wasn't flexible enough. Instead, he used complex Minkowski space, which is also consistent with theories of electromagnetism and quantum mechanics. Because "it is very important that any model constructed to describe psi must not at the same time generate weird or incorrect physics," Targ believes that complex Minkowski space also gives us our best physical model to describe psi phenomenon.

"The complex Minkowski space is a purely geometrical model formulated in terms of space and time coordinates, in which each of the familiar three spatial and one temporal coordinates is expanded by two into their real and imaginary parts making a total of six spatial and two temporal coordinates. There are now three real and three imaginary spatial coordinates, together with the real and imaginary time coordinate."

Particles as we know pop in and out of our reality in the subatomic world. This complex Minkowski structure and gives us an understandable and trackable place for these particles to go to and from when not in our reality.

How does this relate to nonlocality? Nonlocality allows for what studies have consistently demonstrated about clairvoyant reception. After examining a hundred years of data on psi resarch, Targ found there there's no loss of accuracy in any kind of ESP due to distance.

Perhaps this phenomenal reliability and fidelity can be explained by the fact that "desired information is always present and available" when clairvoyants reach back in time. This means that clairvoyants are simply better at accessing what is essentially readily available information. The information's always there; you just have to be open it.I agree with Targ when he says that "psi is often seen as paradoxical because we presently misconstrue the nature of space-time in which we reside. The 'naïve realist' picture of our reality says that we are separate creatures sitting on our own well-circumscribed points in space-time. But for the past thirty years, modern physics has been asserting this model is not correct."

Spiritual and Philosophical Traditions

What we need to begin to explore is how this science intersects with our spiritual and philosophical traditions. How does this science manifest itself in our reality? Or, to ask the question another way, how do religious, spiritual, and philosophical traditions support science?

Targ cites Aldous Huxley, who wrote about the "perennial philosophy," what he called the common thread found in all the world's traditions of religion and wisdom. For Huxley, consciousness is the most important thing in the universe. This isn't a clockwork universe, a great machine; so much as it is like a great thought. We have a dual nature, local and nonlocal, physical and nonmaterial. With our nonlocal, nonmaterial aspect, we can access all of that great thought, the entire universe, and eventually become one with all of it, that is, become one with God.

What Huxley has done is to find the common denominator of all religious and wisdom traditions. Now watch.

When you do the same with science, particularly the continuing evolution of the kind of science we have been discussing, Targ says that "we are experiencing in every area of human activity what Marianne Williamson calls 'a climax in which science and religion are becoming coherent in the exclamation of a single unified truth'."

In short, science and religion are saying the same thing. They are coalescing, not moving apart.

I want to move through the layers of science and get to how we interact with this ever-present accessible information. While we still don't fully understand the mechanics of the science, we can reliably use it and quantify the results. That is what Targ was doing for the government while at Stanford University. In one trial, a subject Targ worked with, a man named Ingo Swann, described and affected a superconducting magnetometer buried under the physics buildings on the Stanford campus. In another instance, during the police investigation of the Patty Hearst abduction, a psychic named Pat Price was able to look through a book of local mug shots, and pick out Donald "Cinque" DeFreeze. As it turned out, DeFreeze had escaped from prison a week earlier, and had been the leader of group that had snatched Hearst. Price followed up by describing the car they'd used and telling the police where they had abandoned it.

So, here's the summary after examining our two examples. As Targ has said, he doesn't just believe in ESP, he has seen it work. Targ is saying that there is a difference between belief and knowledge, and he is unequivocally telling you he knows this works.

Here is the point made in both of the last two chapters: Radin says we should stop funding research to prove psi exists, because we have already proven it. Targ's psi is a given, so what can we do with it?

Let's be precise, even to the point of splitting hairs. We haven't proven the soul, but we have proven an out-of-body consciousness that is common to each one of us. We have proven that prayer works. We are also seeing a convergence/coherence of science and theology that we have never seen in our history. Next, I want to hear from the sensitives themselves and see how they view the means by which they process and receive information.

CHAPTER 13

THE SENSITIVES

Throughout the preceding chapters we have heard from the scientists who have painstakingly proven different aspects of psi research. However, I have not yet discussed those people in our population who are sensitive to this proven human ability. These people have a natural aptitude for psychic ability. We all can learn to play the piano, but some of us are naturally gifted in this area.

There are people in the world who have a natural ability, or some natural talent. A musical prodigy is a good example. We all have music ability at some level, but there are those among us whose musical ability is naturally greater. There was a movie some years ago starring Robert Redford called *The Natural*, who had a unique ability to play baseball. There are those among us who have a natural ability to be sensitive to the energy currents that exist around us.

Very often these people become footnotes to history, anomalies, and sideshow curiosities. Still, what they are able to achieve defies explanation, or defied explanation with the understanding that existed in 1900 or 1930, or even 1960. Today, with these advances in science, we can look back and reexamine some of these people, and note that they are self-similar. That is, they are all similar to one another, an example of fractal geometry raising its head in the real world. Who are these people?

Frederick Bligh Bond did the initial work in the early part of the twentieth century excavating Glastonbury Cathedral. In 1910, he was getting phenomenal and unexplained results by accessing a hidden energy, which he could not even explain to the people he was working for, for fear of losing his job.

The most famous American in this category would undoubtedly be Edgar Cayce. His work is well documented, as is that of his contemporary, Stefan Ossowicki. They did most of their work in the 1920s, 1930s, and 1940s—work that confounded the scientists of their day.

Pat Price and Hella Hammid worked at the Stanford Research Institute along with Russell Targ , but only one, so far as I know, wrote about how he did it and the system he used. That is Ingo Swann, the most notable sensitive (naturally gifted) psychic of the latter part of the twentieth century. Unlike other psychics, Swann wrote about his experiences from the psychic's point of view. He described in detail the process the mind must go through to access the information that exists in wave form around us all. For most of us, our bodies are designed to block this information, with the result that we live below our psychic abilities. But I want you to see this event both from the point of view of the analyzer (the scientist) and of the analyzee (the psychic). Swann believes that knowledge and information exist in some kind of supranormal state, which individuals can tap into via their supernormal senses. Each individual has this ability, but like those with a talent for music, some of us can play naturally while others cannot. Even if all of us cannot sing well, even the worst of us can develop our musical ability. Swann contends that all people can develop their natural ESP.

I need to refer back to the wave/particle duality of subatomic particles. As we've already explained, each quantum particle also manifests itself as a wave. Our bodies are composed of particles. That is how we exist in this reality of three spatial dimensions and one temporal dimension. The particles that compose us, however, have a wave-like ability, and it is this wave-like ability that can access what Swann calls the *psychic information of the universal data banks*. But for us to process that information and bring it into our consciousness and awareness, it must go through what he calls our mind mound. We have an ESP core composed of preconscious processes, a subliminal layer, a conscious layer, our own rooted values, our education, and our beliefs.

It is this composite core of information that can rise up into our conscious mind.

Swann describes these layers of the mind as being like strata in geology. It is better to see this than explain it. This is how Swann represents it. Read it from the bottom up.

ESP emerging in our conscious thoughts

Rooted Values, Preconceptions, Education and Beliefs

Conscious layer

Subliminal Layer

Preconscious processes

The ESP core

The Unconscious

Psychic Information

(Universal Data Banks)

Information travels up from the universal data banks

Before this information can reach our conscious minds, it must travel through all these layers: the unconscious, the preconscious processes, a subliminal layer, a conscious layer, our own rooted values, our education, and our beliefs. Because our rooted values process this psychic information, they can negate or twist our ability to access it.

Swann demonstrates the general flow pattern of psychic information and shows us some of the barriers that impede it. He says that "logic or internalized standards can obliterate the psychic information and often do."

Yet, for all that Swann had to offer, some of the most famous psychic researchers ignored him. When he worked with the great J. B. Rhine, Swann heard from a friend that Rhine had instructed his staff to pretend to be interested, but not to take Swann seriously. This example shows the difficulty in studying ESP, even among some of the best researchers. They refuse to listen to the subjects that are giving them the best results. Swann echoes others when he turns to subatomic physics to explain psi phenomena. Because subatomic particles are also waves, he writes in his book, and waves are not located in a specific place, as a wave, any bit of matter can be "all over the entire universe or at different points at the same time... all a clairvoyant does is bring out, by appropriate focusing, the wave structure of a distance object, which is latently present in any particular point in space."

I don't want to bog you down with all the minutiae, so let's leave it at this: a scientific structure is emerging that explains the phenomena we have seen for centuries but haven't understood.

What does any of this have to do with reincarnation? First, it shows that we really don't understand our own reality. Second, the information about our current lives and all past lives is *out there*, somewhere in the ether, and some can access and communicate with it. In his book, Swann does not show you the results as a scientist would, but instead shows you the path he travels as a psychic. And he says anyone, with training, can travel the same path .

Swann felt that a cynic "who disbelieves something seldom sees facts attesting to the inaccuracy of the disbelief." As everybody knows, people only see what they want to see. This is particularly true when it comes to phenomena that challenge the current metaparadigm. Skeptics will try their best to ignore evidence contrary to their viewpoint, and they'll be quite successful if they keep plugging their ears.

Dean Radin's response to the critics, who say they'll believe it when they see it, is that they've got it backwards. Until they believe it, Radin says, they won't see it.

While I was raising my children, to teach them to rely on their own judgment, I would often tell them something that wasn't true. When they questioned me, I would look at them innocently and say "Who are you going to believe, me or your own eyes?" Then I would explain to them that they were right. By doing this, I was trying to teach them to rely on their own judgment. I believe that this made them the strong, independent young men they are today.

So I say that you cannot believe anything I am telling you, any more than you can believe anything any critic tells you. You must rely on your own eyes, but to do that, you need to distrust them. You have to see this without your own eyes. How many of my readers have heard the phone ring and just knew that it was a sibling, or had a premonition about something that was going to occur? Without seeing it, you believe it because you've experienced it.

So do we see psychic phenomena in the real world and not recognize them for what they are?

Derek Paravicini, who was seen on the *60 Minutes* TV show on March 14, 2010, is an idiot savant.

Born three months premature, blind and disabled, he was unable to communicate. At the age of three, he accidentally passed a room where somebody was teaching music. He rushed in, pushed the teacher aside, and tried to play, karate chopping the keyboard. They tried to pull him away, but amid the noise of him banging on the keyboard, the music teacher heard the melody of "Don't Cry for Me, Argentina," which Derek was trying to play.

The teacher called Derek's parents and asked if he could teach Derek to play. They agreed. When Derek was four his teacher would rush to the keyboard to play a few notes, because Derek would then rush to the keyboard and bang on the keyboard again. One day, it dawned on Derek that the teacher was not trying to take the piano away but trying to show him how to play. At that point, Derek was finally able to communicate.

Derek is now a world-class musician, able to play any song in any style in any key upon request. He has given concerts globally, but away from the keyboard he's virtually unable to function. When Leslie Stahl asked him to hold up three fingers, he was unable to do so. Not only was he unable, but he even asked her how to do it—he didn't know. He doesn't know how old he is, and he'll tell you that. He doesn't know when his birthday is. He cannot read or write. But when he is in front of the keyboard, he is one of the most brilliant pianists you will ever hear.

So why would I mention Derek in a book about reincarnation? Reincarnation is simply a part of the human condition, and to understand this, you have to understand where a person like Derek fits in.

We've already explained that memory isn't located in the human body. Instead, we have to access to memory either clairvoyantly or telepathically, because in our ESP core we are both particles, as exhibited in our human bodies, and wave forms, as (we hypothesize) exhibited in our souls. It can be hypothesized that Derek's ability comes from an ability to access wave forms.

In that case, his disabilities allow him to achieve greater access, which is where the idiot savant form takes its shape. If Derek was cognitively aware, all his analytical skills would bog him down with predetermined mindsets about different things; his reasoning skills, then, would impede the flow of ESP information that is readily available to him. However, he's got a speed pass to that second information stream that Ingo Swann talks about.

In a bittersweet way, because Derek is disabled and lacks preconceptions, labels and myths, education and beliefs, rooted intellectual values, and cultural, educational, and value imprints, he has this unique mastery of music.

So, if we all can access this information to a greater or lesser degree according to our individual abilities, are there other uses for this nonlocal sight? Here is one suggestion.

As part and parcel of our space exploration, I would suggest that sensitives be used to probe the galaxy—that a psi space force be formed. Let it be used to explore our galaxy and look for life. More sensitive and further-reaching than our finest telescopes, these are people whose talent is learnable, that gets better with practice. Let these people be trained. That's one way to use what we learn in the science of reincarnation.

I would suggest a mission to the center of the galaxy with the best psychic minds that we have, in an environment they say would best allow them to make the trip. The same ability can be used to look at subatomic events, as well as locate downed aircraft, as was shown in the CIA operation in Africa that Russell Targ worked on.

If NASA is space exploration, or more correctly, outer-space exploration, then this science and these events merit the same national initiative to explore inner-space events.

Before we leave this chapter there is one more sensitive I would like to talk about, and that is Michael Talbot. Talbot, the author of *The Holographic Universe*, was a gifted science writer. We spoke about him earlier. In 1987 he wrote a book entitled *Your Past Lives: A Reincarnation Handbook*. If Ingo Swann explains how accessing psi works, and its psychic structure, then Talbot gives you a handbook on how to do it yourself.

Talbot was not only a gifted writer, but gifted in his "sight" as well. You can go to the website Thinking Allowed and watch an interview he did with Jeffery Mishlove.

The point of this little passage is that we should use both these books to train other psychics, and they should be tested and examined at The Princeton Anomaly labs.

The military should have its own force of psychics, but they should also have intenders, to influence events at a distance.

Could they look into space to find other intelligent life or the structure of an atom? Could they look into the Russian or Chinese defense command, or as a group try to follow someone who has just died?

As wild as this sounds, there is scientific justification for just these kinds of experiments, experiments that it wouldn't be possible to fund, unless there is a scientific structure to test the results.

This is part and parcel of the science of reincarnation, because any out-of-body manifestation of intelligence makes the concept of consciousness without a body all the more likely. In short, in the idea of nonlocality comes the scientific justification.

It is testable, it submits to math, logic, and replicability. It manifests itself in the macro world in children who claim prior lives, in past-life regression, and in the psychics we have seen in this chapter. It is consistent with the claims of near-death experiences.

It is as important to explore this inner space as it is to explore outer space, if not more so, because in doing so we will find our true nature. This is science, not myth or belief.

CHAPTER 14

PERCEPTION

So, I have a question for the readers of this book. There is something that we, as a race, commonly believe, and it is supported by observation (University of Virginia, near-death experiences, past-life regression), but this scientific theory and clinical proof (clairvoyance and quantum physics) is denied by the same scientists who, when they go to their churches, synagogues, or mosques, believe exactly what they deny in their jobs.

In this entire book, I have stayed away from discussions of religion, and yet the cultural implications of what Michael Newton is saying must be addressed. His subjects speak of an afterlife and then a subsequent incarnation, and it really doesn't matter which religion you follow. Your religion is consistent with his result, so nothing in my book denigrates or is incompatible with any religion, and the science supports it. It doesn't try to explain it. It doesn't say one religion or belief system is better than another. Scientifically, they all seem to work the same.

So even though I don't like Newton's presentation, I must admit that it lines up with the geometry, the other anecdotal information, the cultural information, the structure of the Minkowski Complex, in terms of a new religion, and a new time-space to make what we're seeing work. The Minkowski Complex, by the way, is not just about a space where souls exist. It's about where particles pop in and out of existence. In short, if a subatomic particle disappears and then reappears, as we're seeing in electron microscopes and hadron colliders, where does it go? Why isn't it in our time-space, and if it exists in another time-space, then Minkowski Complex is describing that other space.

Epistemologically speaking, there is no valid scientific proof that anything in this chapter could possibly contribute to a proof of reincarnation.

Except statistically.

If statistics is the great equalizer in determining the probability that one thing is truer than the other, then some other disturbing patterns emerge by looking at data from across a variety of scientific disciplines.

In Newton's body of data, we see very specific descriptions of the afterlife, including a variety and specific number of levels that each soul can pass through, but we see the same thing when other scientists perform past-life regression. Look again at the information that came to Weiss.

The children in the University of Virginia study are less specific about what goes on between lives, but their comments mirror those from the other two groups.

So, then the question becomes whether a mechanism in the human body triggers a common delusion.

You seem to need to choose between the following two possibilities. Either this is a common delusion created by our bioelectrical postulates, or we really are seeing some version of an afterlife.

If you then couple this with the other factual evidence—for instance, that children who claim prior experiences are accepted by their prior families and that clairvoyants are able to provide the location of material objects that none of us could possibly know—then we may be seeing a fairly consistent version of what our individual futures are.

The interesting thing about this construct is that your intention to create it seems to make it so. We can say this with a reliable deviation of 6 to 8 percent. This is the difference we find between the figures in the Princeton University study. In that study one person can make an image in a computer show up 52 percent of the time, and the other image show up 48 percent of the time with their intention alone. A bonded pair of individuals, a couple, can make that percentage deviate 8 percent, or the difference between 46 percent and 54 percent.

If, then, 8 percent of us create this afterlife with our intention alone, what just happened? Did I just say you create the afterlife by your intent? Newton describes an afterlife that the creator, whoever or whatever that may be, has created—an environment from his/her own intent. Do we have the same nascent ability? Can we be co-creators by using physics, on which the Intention Experiments are based?

I want you to be offended at such a rash speculation, for that will allow me to introduce John Craig Venter.

Did I just say that by intention you create the afterlife? You would probably say no. If I said that with a computer program and chemicals I could create life, would you also say no? If you did you would be wrong.

John Craig Venter, an American biologist and entrepreneur, is most famous for his leading role in sequencing the human genome, and for his role in creating the first cell with a synthetic genome, in 2010. Venter, one of the leading scientists of the twenty-first century, founded Celera Genomics, the Institute for Genomic Research, and the J. Craig Venter Institute (JCVI), and is now working to create synthetic biological organisms.

With a computer program and chemicals, Venter created self-replicating life. He programmed a DNA code for a self-replicating organism.

Genomic science has greatly enhanced our understanding of the biological world. It is enabling researchers to *read* the genetic code of organisms from all branches of life by sequencing the four letters that make up DNA. Sequencing genomes has now become routine, giving rise to thousands of genomes in the public databases. In essence, scientists are digitizing biology by converting the series of A, C, T, and G that make up DNA into ones and zeros in a computer. But can the process be reversed? Can you start with zeros and ones in a computer and define the characteristics of a living cell? Scientists set out to answer this question. Using the binary code of the computer, they tried to design a chemical makeup of something that would live.

In the field of chemistry, once chemists determine the structure of a new compound, they then synthesize the chemical to determine if the synthetic structure functioned like the original compound. Now, Venter and his colleagues are doing the same with genomes. In 2003, they created a virus; in 2008, they synthesized a small bacterial genome; and in 2010, the synthetic genome was used to create the first cell controlled completely by a synthetic genome.

Twenty-five years ago, you could not conceive that a scientist would create life by using four chemical compounds and a computer, but that's what happened.

Now here's the point. Whether you agree or disagree with the finer points of whether Venter has actually created life, the people with the money—the real money—believes he did it, because he's now worth hundreds of millions of dollars. And if we can create life from chemistry, we can create life from physics. That is how you create the land of the soul—intention. So, if we can create life using a computer and chemistry, as Venter did, then why would you be offended by my speculation that we can do the same thing using physics? This brings me to why I am angry about tumbleweeds.

For any American, and especially Americans between the ages of forty and ninety, tumbleweeds are as iconic an image of the American West as the buffalo or the Indian. Hollywood movies have exported that image around the world. When Jimmy Stewart walked down the street to fight Liberty Valance, tumbleweeds tumbled down the street. Every movie cowboy, from Tom Mix and Hopalong Cassidy through John Wayne, that walked down a dusty western street or rode across a high desert, had tumbleweeds blown across his path. When I traveled west, as I frequently did, the first plant I wanted to see was tumbleweed. I wanted to see them rolling along.

"See them tumbling down, wearing my heart as a frown. There on the range I belong rolling along with the tumbling tumbleweeds," sang the Sons of the Pioneers and Gene Autry. It was, frankly, an image I loved.

The problem is that the image is not true. There never was any tumbleweed in the West during pioneer days. According to Wikipedia, tumbleweeds were introduced into America in the 1870s, when Russian farmers brought flax to America to plant. During the Dust Bowl of the 1930s, the U. S. government also introduced flax imported from Russia. Flax could grow in dry, arid soil, and they hoped that planting it throughout the Southwest would control the massive erosion that occurred during the 1930s. They accidentally introduced a weed, because the plantings contained tumbleweed seeds that grew naturally in western Russia, but had now migrated to the United States. So, the flax seeds contained tumbleweed seeds. By the late 1930s and mid-1940s, the cowboys were singing about them. (Interestingly, tumbleweeds like to live in areas of uranium deposits, but this piece of information has nothing really to do with the story.) This is the important point—there were no tumbleweeds in the United States before 1870. It would probably take them ten years to colonize the entire West. So despite Hollywood's depictions, tumbleweeds did not really exist in the West until 1885 to 1890 at the earliest.

In 1893 Frederick Jackson Turner advanced the Frontier Thesis, saying that America's frontier was coming to an end. The U.S. Census of 1890 had officially stated that the American frontier had broken up. Both these sources were saying that Americans had pushed all the way to the west coast and that there was no more frontier. The West as Americans had known it had changed. Hollywood depicted the West as it was seen in the 1930's onward, a frontier with tumbleweeds. But as people pushed west through the period of Lewis and Clark (1804–1806), the mountain men, (1810–1840), the Mexican-American war (1848), the California gold rush (1849), the post–Civil War cattle drives (1860s), and the driving of the golden spike to finish the first transcontinental railroad (1869), there was not even one tumbleweed.

This story is about perception—the fact that reality has a nasty way of intruding on those images and concepts we like, enjoy, and hold dear. Sometimes we do not want to accept that new reality, but it lies there like a Zen puzzle, waiting for us. I don't like it when reality intrudes on my nicely structured image, and neither do most people. You may not accept the science presented here for that reason. But that mental resistance to epic change seems to be common. Here is another example.

Alfred Wegener was born in 1880; in 1904, he received a Ph.D. in astronomy from the University of Berlin. He was a prominent meteorologist, and one of the first to use balloons to track air currents. The book he wrote on the subject became the standard text in Germany, and he taught at universities in both Germany and Austria.

Wegener published his book describing his theory of continental drift, *The Origin of Continents and Oceans*, in 1915. It was almost universally rejected by academics and the scientific establishment, and in the harshest terms. One critic said that "Wegener's hypothesis in general is of the footloose type, in that it takes considerable liberty with our globe and is less bound by restrictions, or tied down by awkward and ugly facts than most of his rival theories."

What Wegener proposed was that about three hundred million years ago, there was a single large land mass on earth, which he called Pangaea, "all earth" in Greek. Pangaea had broken up and the pieces had drifted apart, to form the continents we know.

"Ridiculed" is a mild term for his critics' response. Wegener died discredited and broken. Wegener's problem was that he had no convincing mechanism to explain how the continents could move. He thought the continents moved through the earth's crust like a plow moves through a field, but he had no way of proving it, and if he'd thought of one, it would have proved him wrong. The actual mechanism, what we know as plate tectonics, was widely accepted by geologists and the scientific establishment by the 1960s. Wegener got the mechanism wrong, but his central insight proved correct, and his theory of continental drift is now part of our understanding of the earth.

Now, there are several points that I want to draw from this. First, as little as one hundred years ago, our understanding of plate tectonics didn't exist, and when Wegener first proposed the continental drift theory, which was the basis of the theory of plate tectonics, he was lambasted.

Our case for reincarnation is getting some of this same rejection. Just as Wegener was an interdisciplinary scientist, our case here is an interdisciplinary one, based in archeology, geology, chemistry, biology, physics, and more. The current facts are lining up in a certain way, pointing to a certain conclusion. Just as Wegener saw South America and Africa fitting together without being able to determine how it worked, so the pieces of the puzzle of the science of reincarnation are fitting together, though we don't yet have all the facts. What will this puzzle look like one hundred years from now?

So here is the point of all this, Wegener looked at the observable data and came to the correct conclusion (continental drift/plate tectonics) but did not know what caused the movement of the continents. It took around fifty years for that science to emerge.

We don't know what mechanisms cause people to remember past lives when they go through past-life regression, but it is a common human experience and the stories about life between lives are too consistent to be fabricated by so many people who don't know each other.

I give you the story of Wegener here so that you will have a context about the nature of scientific discovery and the development of ground-breaking ideas.

But this book, *The Science of Reincarnation*, is about facts that challenge our perceptions. I did not want to give up my perceptions about the Old West, images I had grown up with and had affection for. The images were factually wrong, shaped by a Hollywood movie and television industry that I enjoyed. But if I am going to base my view of reality on the facts, then I must acknowledge that there weren't any tumbleweeds in the Old West.

I want to add one last point about frontiers. In the opening of every early *Star Trek* episode, Captain James T. Kirk of the *USS Enterprise* called space the final frontier. I disagree. Today we stand on the greatest frontier mankind has ever known. That frontier is ourselves.

As we move to the next chapter and toward the conclusion of this book, I want you to leave your perceptions here with the tumbleweeds and go solely on the facts. You will have a better experience if you do.

CHAPTER 15

THE NEW METAPARADIGM: DO WE REINCARNATE?

We will discuss two topics in this chapter. The first is whether, based on the science we have presented, reincarnation can be considered a science. The second is whether, based on that science, it is probable that humans reincarnate. The last thing to remember in defining a science is that it defines itself. Our understanding of the science in never complete, so it evolves. So this is not the end point, simply the point where we have coalesced the view, and now each of us has to determine whether, based on this information, the scientific study of reincarnation should be considered a science.

This book is about the science of reincarnation. Does it conclusively make the case that the study of reincarnation should be considered a science? Are we, because of recent discoveries, changing from belief in a multitude of religions that promise an afterlife, to a single fact-based paradigm that proves consciousness continues outside a person's bodily form? Finally, what reliable, accepted scientific method can determine if we can actually call reincarnation, and the study of it, a science?

As I stated in chapter 4, *The Structure of Scientific Revolutions* by Thomas Kuhn is regarded as one of the twenty most important science books ever written; it is in the same category as Newton's *Principia* or Einstein's papers on relativity. Kuhn lays out a hierarchy of science that Schwartz describes concisely in *The Secret Vaults of Time*:

> It should now be clear that there is an entire hierarchy of science within Kuhn's model, one that begins with the individual researcher, goes to the school (sometimes literarily the institution with which the researcher is affiliated); then to a discipline; then a paradigm-achieved discipline (or science); finally a multiscience community made up of the disciplines that have achieved paradigm and share in the metaparadigm.

For the study of reincarnation to be called a science, it has to meet not only Kuhn's categorizations, but globally respected research institutions should also be pursuing courses of independent study that are subject to replicability and peer review; in short, it should follow the standard research methodology accepted by the scientific community.

I believe this is already a fact.

This is an epic scientific advancement, because it means is that the metaparadigm by which we view our reality is changing. Such a change has occurred twice in the last two thousand years; this is the third change. Here is a quick review.

The Genesis Metaparadigm

The first metaparadigm was the Genesis metaparadigm. All science until the enlightenment was viewed through the lens of God creating the heavens and the earth. Galileo was sentenced to house arrest because his observations of the planets contradicted the church's teaching.

The Grand Material Metaparadigm

The current metaparadigm that is being phased out, the *Grand Material Metaparadigm*, has several critical assumptions. What follows is a composite understanding of this paradigm.

1. Each consciousness is a discreet entity, man is essentially isolated from the world, and his mind is isolated from his body.

2. Organic evolution operates based on Darwinian natural selection. Man is a survival machine powered by chemicals and genetic coding.

3. The mind is the result of chemical and bioelectrical processes. The brain is the home of consciousness and that

consciousness is discrete. It is created and driven by the chemistry and bioelectricity of the body.

4. There is only one time-space continuum, and it provides for only one reality, and that time and space are finite. The Universe is made up of particles and quanta that are separate from one another unless connections are made through fields.

5. Nothing travels faster than the speed of light.

The New Metaparadigm

"It is apparent that a metaparadigm crisis is approaching, although its development is still in the early stages.... But it is certain that the Unified Metaparadign (for that is a reasonable name) will no longer maintain the artificial distinctions between consciousness and the physical, time and space, or researcher and experiment. Rather than dealing with compartments, it will see a continuum, a spectrum running from physical reality to suprasensible realms—as visible shades run to X-rays and beyond."

—Stephan Schwartz

Schwartz calls the new metaparadigm the *Unified Metaparadigm*, because it will no longer maintain the artificial distinction between consciousness and the physical. Man is a being of transcendental consciousness. I want to define *transcendent* here. According to Webster's Dictionary it means, "extending or lying beyond the limits of ordinary experience, transcending the universe or material experience."

We have proven a disembodied form of consciousness in archeology, in the Intention Experiments at Princeton University, and in many other corollary experiments. We have cited Deepak Chopra explaining that consciousness is more permanent than your body. Thus, instead of compartmentalizing each human experience, we should see life as a continuum, and then life after death will match the unfolding metaparadigm.

What follows are the key points for the new emerging metapardigm, as described by Lynne McTaggert.

1. The universe is a single entity in which every part is connected to every other part.

2. The communication of the world did not occur in the visible realm of Newton, but in the subatomic world of Heisenberg.

3. Cells and DNA communicate through frequencies.

4. The brain perceives and makes its own record of the world in pulsating waves.

5. A substructure underpins the universe, which is essentially a recording medium of everything, providing a means for everything to communicate with everything else.

6. People are indivisible from their environment. Living consciousness is not an isolated entity. It increases order in the rest of the world.

What the scientists are telling us is that you cannot separate the experiment from the observer, that everything is a wave form that we interpret through our senses as this reality. This emerging construct of our universe in itself supports the theory of reincarnation.

Radin's Version of the Unified Metaparadigm

While Schwartz and McTaggart generalize about the coming change to the metaparadigm, Radin is more specific and develops an actual theory. He says, "It turns out that some scientific developments in recent years suggest a way of thinking about psi that is also compatible with mainstream scientific models. Four such developments are related to quantum theory. All four run counter to common sense . . . and all four have now been empirically proven. . . . [The] first, not surprisingly, is the idea of nonlocality itself. The second is that quantum effects may be important to consciousness and biological organisms. The third is that information can be transmitted without expending energy. And the fourth is that information can be instantaneously transmitted—the actual word that physicists use is 'teleported'—from one place to another, independent of distance."

I believe the second should be worded that quantum effects are absolutely important to consciousness in biological organisms.

Radin says that "our most sophisticated scientific theories about the way the world works have not caught up yet with these phenomena." What we need to do, therefore, is incorporate the discoveries of the quantum world into biology, and then we will find a basis for a theory that deals with the anomalies we are currently seeing in the sciences.

I have cited three respected scientists and their views. There are more out there, but these views are representative of the unfolding, new world view.

Summation

Should reincarnation, then, be considered a science in light of the changing metaparadigm and the scientific discoveries of recent years?

What We Have

Anecdotal Information/Field Observations

This would include the studies done at the University of Virginia. Children who remember prior life experiences are global phenomena.

Near-death experiences are also a global phenomenon. The reports are amazingly consistent.

Past-life regression also falls into this category.

Although there is no proof in this category, two things stand out. The first is that all who report such experiences describe them consistently. The second is that while they cannot be explained by the Grand Material Metaparadigm, they fall neatly into the theory of the emerging Unified Metapardigm.

Clairvoyant Archaeology/Psi

Clairvoyance is now considered proven, because of results from the science of archeology. We now have an accepted case of a disembodied form of consciousness that reaches through time and space. It is supported by both the emerging metaparadigm and anecdotal information from field observations.

This ability has not only been proven conclusively in the Intention Experiments at Princeton University, but those results have been confirmed by other labs and experiments run globally. Every human being has such an ability to a greater or lesser degree and can develop that ability through their own individual efforts. This ability is called *psi*.

Through the meta-analysis of psi we have hundreds of millions of trials performed all over the world in studies covering more than 140 years. I think it is reliable to conclude that psi exists, based upon this astounding degree of replicability.

Complex Minkowski Space

For the very first time in human history, we have a spatial and temporal construct. Some models vary, but for the understanding of the layman, it is enough to note that over the last twenty-five years eight to eleven different dimensions have been theorized. This is an evolving model, and the number may change again. But all the models agree that our reality is three and a half of them. We have width, height, and depth; time is the half, because we can only move forward in it. All dimensions are totalities, so all of time exists.

The experiments in quantum physics confirm the existence of the dimensions. We are made up of particles that don't exhibit all of themselves in our reality in our time/space. This supports the form of a continuous, disembodied consciousness, or a soul, if you want to call it that.

We have a scientific model grounded in physics and experimental results that accommodate a location for a human consciousness independent of the body, which is able to transcend human death. In short, we have a scientific model with a definable location where the soul exists after death.

We're not talking about philosophical models here. We're not talking about religious models either, involving heaven, hell, or purgatory. We're talking about spatial and geometric coordinates that are connected to the human body on the temporal plane through nonlocality. Nonlocality, which we have known about for a century and have proven in the laboratory, could be, as Russell Targ states in *Limitless Mind*, the most profound discovery in all of science.

Quantum Physics

We now have a scientific theory that explains how this phenomenon works. The wave/particle duality that makes up our very atoms also gives us the ability to receive and transmit information in a wave form. This ability to be conscious in a wave form hints at our ability to retain our consciousness after our body's demise.

This understanding of the mechanism is based on the subatomic nature of the very energy we are composed of.

This nonlocal nature is the mechanism that psi operates on, and psi is the manifestation of the nature of energy. It is shown to operate on the following points:

Nonlocality

Duality of particle physics

Energy transfer

Zero-point energy

We cannot see 94 to 96 percent of the matter in the universe

You have heard Dean Radin say that cosmologists have missed 96 percent of the matter in the universe. Lynne McTaggart also put forth the number of 96 percent. So how insightful can our critics be when we have this enormous blind spot involving 96 percent of everything, ever? It seems to me that they're just as clueless as we are, if a bit more outspoken.

Let's be clear for all of us. We only see 4% of the known matter in the universe. Of the theorized 8 to 11 dimensions our reality is only 3 and a half dimensions, 3 spatial dimensions, and we only move forward in the one temporal dimension and I count that as a half a dimension. So our inferring the whole from the limited sample of evidence we have can include things we have not yet imagined. What we can do is be open to what we do have is telling us.

Quantum Biology/Digital Biology

For the very first time, we have a theory that transcends biology and deals directly with physics and the human body.

If particles in our reality are not all here, neither are we. We see this anecdotally in Michael Newton's descriptions, although he never mentions particle physics.

Evidence has been found that electrons can jump into and out of the realm humans reside in, which, along with quantum theory and the Minkowski space, demonstrates that this world and this realm are not the only planes of existence for the human intellect.

The testimonies of Dr. Newton's patients about life after life supply additional evidence for the science of reincarnation. The various stories narrated about these experiences are so consistent that they further strengthen the case for reincarnation: a case that grows stronger with each successfully replicated answer.

Nonlocal in nature means we have an aspect to us that is not local, meaning that what is here is not all of us. The particles we are made of pop into and out of our time-space. The very smallest are not particles at all, but energy fields. If the soul is to exist without a body it must be made of energy. For the very first time, therefore, we have a science to give a theory to the anecdotal information that Dr. Newton has given us.

Finally, we have consistency of theory and effect. The best theory is that the nonlocality we see in the subatomic world and the duality seen in particle physics are both caused by the same phenomena we see in the macro world in clairvoyance and telepathy. It makes sense, even if we don't have final proof. If it has webbed feet and quacks, it's probably a duck.

The human body regenerates itself every two years, completely replacing every molecule and atom in the body. The stomach lining replaces itself once a day. Skin replaces itself every six weeks. Even the enamel in teeth replaces itself every two years. The person that existed two years ago is completely different from the one that exists today.

But how does our memory stay intact if our body is constantly renovating? For example, how do we remember the taste of strawberry ice cream? That memory, we now know, is not housed in our body, but is accessed by our body. Therefore, what is here is not all of us, which strangely coincides with Dr. Newton's descriptions of life between lives.

Fractal Geometry

Until recently, the insights of fractal geometry were not available. Fractal geometry is built on self-similarity and iteration, both of which exist throughout the natural world and both of which support reincarnation. The self-iteration is life-after-life-after-life, the geometry is the mathematics of quantum nature, and the self-similarity can be said to be represented in the life-after-life we're seeing the children lead in Tucker's studies, but also in the conscious aspect of the mind.

The complex Minkowski space that Targ discusses in his book gives an eight-dimensional view of reality; that is, six spatial dimensions and two dimensions of time.

At some point, we must reconcile our conception of reality with the understanding of quantum mechanics. In order for the mathematics to work in quantum mechanics, I have heard that as many as eleven dimensions are needed and that many of them are quantum, meaning they are so small that they have no dimensional structure that could be easily understood. The point is that we are beginning to reconcile a common understanding with what we're seeing in the quantum world.

Statistics

So how do I make all of this real for you? I'll use a branch of mathematics that we can all trust—statistics.

Even the hicks from the sticks know that in baseball, a lifetime batting average of .300 is pretty good; it means that the player will get three hits in ten. If we apply statistics, as the scientists have, and come up with a greater number of positives than negatives, consistently, through every test, we can safely accept that the positives are what science is trying to tell us. If we get a different result, then science is telling us something different.

So if we line up the anecdotal events—the children who claim prior lives, the people who have near-death experiences, those people who experience past-life regression—all of them relate a similar experience. The odds against all those tens of thousands, if not hundreds of thousands, of people relating similar stories is an astronomical number to one.

Psi is as well established as any other statistical phenomenon, and the reality of reincarnation has been established as well as any other statistical phenomenon as well.

Only one thing stands between you and acceptance—your presuppositions, but these are a huge thing to overcome. Until they're overcome, we can't objectively pursue the path of logic that lies before us.

Statistically, reincarnation should be accepted as well as psi is accepted as statistical phenomena. The odds against the hundreds of thousands of near-death experiences, prior life claims and past-life regressions all following the same pattern are astronomical. While the theories about the causes of these events are still evolving, we can accept the effects we are seeing as a reliable representation of reality.

From a sociological view, the statistical aberration—the finding from mathematical analysis—that all of these different and disparate groups would tell a similar story is disturbing. It's disturbing, because the statistical value that this would happen accidentally is a one with a lot of zeros after it. In short, it is highly unlikely to occur.

But when I join it with the hard data that we can influence machines with simple intent, that prayer has been shown scientifically to be effective, that memory has a model of holographic storage, that clairvoyance produces results, and that a theory founded on subatomic or quantum physics shows how the mind can reach outside the body, all this leads us to believe that this scenario may be probable and that its description of an afterlife is true.

The Structure of Scientific Revolutions

If this book does anything, it shifts the subject of reincarnation from the world of the paranormal, occult, religion, and belief into the world of hard science. It is a subject to be studied. This science incorporates field observations, theory, a mathematical model, and experimental proof for a form of disembodied consciousness. When all this is considered, regarding reincarnation as a science is justified.

This is because it meets all the criteria for a science as defined by Kuhn. It has a paradigm with a mathematically based formula that predicts future outcomes based on past performance.

In the future, this mathematics will be applied to the data at the University of Virginia, much as it took Johannes Kepler to interpret Tycho Brahe's observations of the movements of the planets and stars. When this happens, we will reorder the heavens.

In the meantime, like Wegener, without all the pieces we can only speculate about future outcomes. But we can't deny that Africa fits neatly into South America, any more than we can deny what we are seeing in the science of reincarnation.

Who Believes This?

From 1981 to 1985, five different government-sponsored scientific review committees were given the task of examining the evidence of psi effects, and their conclusions all supported psychic phenomena. All of the following facts come from Dean Radin's book.

"In 1985, a report prepared for the Army Research Institute concluded that the data reviewed constituted genuine scientific anomalies."

"In 1987, the National Research Council said this work should be taken seriously."

"In 1989, the Office of Technology Assessment issued a report of a workshop on the status of parapsychology, which concluded that the field of parapsychology appears to merit consideration."

In 1995, the American Institutes for Research reviewed classified government-sponsored psi research for the CIA and concluded that "[the] statistical results of the studies examined are far beyond what is expected by chance."

"The evidence for these basic phenomena is so well-established that most psi researchers today no longer conduct proof-oriented experiments. Instead, they focus largely on *process*-oriented questions like, 'What influences psi performance?' and 'How does it work?'"

So, who would believe this? The U.S. Government does.

And then lies about it. What do I mean by that?

According to Dean Radin, in his book *The Conscious Universe*, "In 1984 the U.S. Army Research Institute asked the premier scientific body in the United States, the National Academy of Sciences, to evaluate a variety of training techniques and claims about enhanced human performance. These techniques included…parapsychology. The National Academy of Sciences directed its principal operating agency, the National Research Council (NRC) to form a committee to examine the scientific evidence in this area.

"In December 1987 the NRC announced its conclusions. 'The committee finds no scientific justification from research conducted over a period of 130 years for the existence of parapsychological phenomena.'" Yet it later came to light that John Swets, the NRC chairman, had asked Robert Rosenthal of Harvard University to withdraw his conclusions from the report because they were favorable to parapsychology. This was clearly revealed when a newspaper reporter for *The Chronicle of Higher Education* asked NRC committee chairman Swets why conclusions were withdrawn. Swets replied: "We thought the quality of our analysis was better, and we didn't see much point in putting out mixed signals." He further explained, "I didn't feel we were obligated to represent every point of view." This meant the NRC committee in effect had created a "file drawer" of ignored positive studies that it didn't wish to talk about.

"There is no need to belabor the point; it is clear that abject prejudice exists in science just as it does in other human endeavors. We were able to detect it fairly easily in the case of the NRC report by comparing the committee's public pronouncements with what the report actually says."

Now I want you to refer back to the timeline of Targ's studies at SRI that I described in Chapter 11. I am not looking for conspiracies, nor am I claiming some mischief was done by the U.S. Government. What I will say is this. If I had been in the CIA or NSA and knew a report favorable to psi operations was going to be published and would perhaps show our research—successful research—to our enemies, I would have had someone suppress and misdirect the information coming out of the NRC.

I want you to remember that the information in this book is all in the public record. It does more than expand our reality; it reshapes it beyond current recognition. Any research coming out now or in the future needs to be fully vetted. Some of it can be wrong, some of it can be bogus, and some may be correct information that will not be accepted because it is beyond what some scientists can accept. Additionally, some information will be suppressed and falsified to misdirect for politically expedient reasons.

But what has been proven and accepted by our scientific community reshapes our entire reality. The words "the coming paradigm change" are simply insufficient to emotionally process this change.

The Russians have their own program on psi. Our government uses it, and it seems that at least on some level we have a psi-science arms race.

So then, do we reincarnate?

We have the following points to consider:

Anecdotal information for reincarnation is statistically consistent to the point of excluding an alternative. In short, a meta-analysis of the globally consistent reports of near-death experiences, past-life regressions, and claims of prior incarnations is so statistically overwhelming that it makes any other conclusion unlikely.

Consciousness remains discrete outside our bodies.

It has been proven that clairvoyance can reach out of the body and extract information. The fact that our intentions can influence physical and nonphysical systems has been proven.

It has also been proven that consciousness is permanent, though your physical body is not. This claim is not made in the context of dying, but because your body recreates itself every two years. In short, every cell that was in your body two years ago has been replaced. So your body is not permanent, but your consciousness is.

It is a fact that your memory, a part of your consciousness, is not stored anywhere in your body, but rather everywhere in your body in a holographic form. Even more, your consciousness may be stored in part of the zero-point field, because it is able to travel in that field outside your body, extracting information and influencing systems outside your mind.

At this point, time does not act as all the other dimensions do, because we can see and travel in all directions in the spatial dimensions but can only move forward in time. It's as though in order to *live*, we restrict time so it does not behave as the other three dimensions do, but being able to move both ways in time is consistent with theoretical physics.

But there is a predictor that we reincarnate, which is indisputable, and that is the paradigm of the science of reincarnation itself. By making the case for the science of reincarnation, we have taken Kuhn's definition of a science and created the strongest case possible that we indeed reincarnate.

The mathematics of reincarnation, fractal geometry, is the mathematics of nature. We see self-similarity and iteration throughout nature. It is that same mathematics at work in the stories of the children studied at the University of Virginia. It is the same mathematics at work in the past-life regression studies done by multiple scientists.

A scientific paradigm explains the science and predicts what is to come. The paradigm of this science predicts that in the future, as we learn more and more about quantum theory and other disciplines within this science, we will learn more and more about ourselves until this will be accepted fact. When that occurs, mankind will undergo a metamorphosis and we will enter a new age, not just of understanding, but of action.

So then, based on the evidentiary data, we can conclude several things.

First, reincarnation can now be considered a science. This is an epic change in human perception about our own reality. This change is not based on any belief system, but on the criteria set forth in the structure of scientific revolutions and on the criteria that it has its own paradigm. There are field observations, theory, experimental data, and a mathematical foundation, fractal geometry, the mathematics of nature that predicts where the science is going. In short reincarnation now meets the definition of what it means to be a science.

Convinced

Now, at the beginning of this book I said I was going to prove that reincarnation should be a science and that I would convince you that you have lived before. I believe that I have made the case for reincarnation to be a science, but have I convinced *you* that you have lived before?

You have heard two respected scientists, Dean Radin and Russell Targ, both say that psi exists beyond any doubt. Dean even calculated the odds against chance for us. Psi, the human ability for the mind to reach out and exist outside the body is manifested in children who claim to have lived before, past-life regression, and near-death experiences. It has proven itself in labs at Princeton and Stanford. But has it proven itself to you? Are you convinced you have lived before based on all this evidence?

If not, the answer may not lie in the information but your own individual belief system creating your own reality. Depak Copra feels that the only thing that is true is what we tell ourselves is true. I had one of my early editors tell me that he did not believe in reincarnation but that he would be resurrected. He could not process this science because he could not get past his own worldview, which was instilled in him early in his life, like the cats Chopra talks about. Their physical view of reality was created as they opened their eyes to the world. That shaped what they could see for the rest of their lives.

The answer for people who are not convinced that they have reincarnated before may at the end of the day lie with Kuhn. When true scientific revolutions come, it takes a generational change. The scientists of the last paradigm need to die off before a new group of scientists grows up in a world where this new information exists. It is only when the generations change that the new generation can accept a radical change in the scientific view.

Therefore it is understandable that some readers will not be convinced that they have reincarnated before, just as scientists could not accept that Wegener's theories of continental drift were correct. But that leaves the rest of us.

I would say it is statistically probable that we reincarnate. I conclude this because consciousness is what is permanent about us, not our body; because memory is stored within us in an interference pattern holographically; because it is proven that our minds can reach outside our bodies through time and space to extract or send information; because underneath all this macro evidence is a scientific theory resting on observations in the quantum world that has met the gold standard of science—replicability; and because we see subatomic particles popping into and out of our reality. For over a century, this probability theory has been a cornerstone of our science.

This book, by proving reincarnation, will redefine the nature of human consciousness. The case for reincarnation is built by the reality of clairvoyance—which has been proven through archaeology. We learned that it has been shown at Princeton University that human consciousness can influence computers simply through intent. We have looked at the anecdotal information gleaned from studies at the University of Virginia that indicate consciousness retains self-reference after death and rebirth. One of the more stunning comments, to me, came from a woman in England, who when asked if she now believes in reincarnation, said no—she's experienced it.

We also see evidence for my proposition in the cultural record of all religions, where it is referred to as life after death. It really should be called *consciousness after death*.

We see evidence in the very structures that we are as human beings—that our physical beings are different day-to-day and year-to-year, yet our memory remains the same. My view is consistent with the geometry of life, that self-iteration that we find in the fractal geometry that we didn't have until the 1970s, when supercomputers could do the calculations necessary to create fractal geometry.

We see evidence in the way physical machines operate: any scientist can trace how an electrical impulse can move from your brain, down your arm, and make your hand move, but no scientist can tell you how the nonmaterial thought becomes the material action. On a quantum biological level, we see that something is happening outside of our bodies to create this.

We see how our memory does not reside in our bodies, but is stored holographically throughout it. In the coming years, we may discover that the whole hologram simply accesses a river of information outside our bodies, and, if that's the case, then that's where consciousness is.

But a conscious wave form may not be able to experience the taste of an orange, or the smell of a flower, or the touch of another person unless it incarnates, and that may be part of the reason why we're here—to interact, to feel, to experience . . . and maybe, of course, to learn a few lessons for our next few lives.

Question: if such consciousness existed and it existed eternally, would it want to limit its opportunities to experience love, fear, excitement, death, and all the emotions of a human condition? Would it want, having lived once, to want to live again? Would you intend to live again? If you did how would you want to do it? Just asking the question, and you the reader creating the intent, creates the possibility.

The End

EPILOGUE

What all this talk of science neglects is the word hope. Most people in the world believe in God and an afterlife. This science supports that belief. It doesn't really matter what religion you are or if you are just spiritual. You have incarnated once, if you are reading these words you are alive now. Living the first time is much more unbelievable than doing it twice. Some people believe in resurrection not reincarnation, this science supports the idea that you will live on after your death even if you don't reincarnate. It is science that is more epic than the exploration of the universe, it is exploration of ourselves.

The ramifications of this science are a book unto itself. It reaches into the social, political and religious fabric of our global structure. There is a tendency to want to avert your eyes and deny it. The thought I want to leave you with is do not be afraid. What leaches out of this information is love, trust and hope. Science is not the new God; it is just the mechanism by which we learn about the universe and ourselves.

BIBLIOGRAPHY

Chopra, Deepak. *Quantum Healing: Exploring the Frontiers of Mind/Body Medicine.* New York: Bantam, 1990.

Kuhn, Thomas. *The Structure of Scientific Revolutions.* 3rd ed. Chicago: University of Chicago, 1996.

Mandelbrot, Benoit. *The Fractal Geometry of Nature.* New York: Times, 1983.

McTaggart, Lynne. *The Field: The Quest for the Secret Force of the Universe.* New York: Harper, 2003.

Moody, Raymond. *Life After Life: The Investigation of a Phenomenon--Survival of Bodily Death.* New York: HarperOne, 2001.

Newton, Michael. *Journey of Souls: Case Studies of Life Between Lives.* 1st ed. Woodbury: Llewellyn, 1994 Newton, Michael. *Life Between Lives: Hypnotherapy for Spiritual Regression.* Woodbury: Llewellyn, 2004.

Newton, Michael. *Destiny of Souls: New Case Studies of Life Between Lives.* 2 sub ed. Llewellyn, 2000.

Pilkington, Mark. Far Out: 101 Strange Tales From Science's Outer Edge. New York: The Disinformation Company Ltd. 2007

Radin, Dean. *The Conscious Universe: The Scientific Truth of Psychic Phenomena.* New York: HarperOne, 1997.

Radin, Dean. *Entangled Minds: Extrasensory Experiences in a Quantum Reality.* New York: Paraview Pocket Books, 2006.

Schwartz, Stephen. *The Secret Vaults of Time: Psychic Archaeology and the Quest for Man's Beginnings (Studies in Consciousness).* Massachusetts: Hampton Roads, 1978

Swann, Ingo. *Natural ESP.* New York: Bantam, 1987.

Talbot, Michael. *The Holographic Universe.* New York: Harper Perennial, 1992.

Targ, Russell. *Limitless Mind: A Guide to Remote Viewing and Transformation of Consciousness.* San Francisco: New World Library, 2004.

Tucker, Jim. *Life Before Life: A Scientific Investigation of Children's Memories of Previous Lives.* New York: St. Martin's, 2005.

Weiss, Brian. *Many Lives, Many Masters: The True Story of a Prominent Psychiatrist, His Young Patient, and the Past-Life Therapy That Changed Both Their Lives.* New York: Fireside, 1988.

Wegener, Alfred. *The Origins of Continents and Oceans.* 4th ed. New York: Dover, 1966.

Red Front Desk
Roof 1113 Butterfield
Inn   D6 DC
         60515